DOGTOWN ECOVILLAGE: GREEN BOOK

EDITED BY AMY HEREFORD

Publisher: RLC, 6400 Minnesota Ave. St. Louis, MO 63111
Web Site: www.ahereford.org/ecovillage

Copyright © 2018, 2025

All Rights Reserved. No part of this publication may be reproduced, stored in a retrieval system or transmitted in any form by any means without prior permission of the copyright owner. Enquires Should be Made to the Publisher.

Every effort has been made to ensure that this book is free from error or omissions. However, the Publisher, the Author, the Editor or their respective employees or agents, shall not accept responsibility for injury, loss or damage as a result of material in this book whether or not such injury, loss or damage is in any way due to any negligent act or omission, breach of duty or default on the part of the Publisher, the Author, the Editor or their respective employees or agents.

Acknowledgements: We acknowledge the invaluable advice and assistance given by the Ecovillagers who contributed their knowledge and expertise to this project. We also thank the Dogtown Ecovillage, the Saint Louis Ecovillage Network and all those who have participated in our community, in our events and in our projects.

Title: Dogtown Ecovillage: Green Book, 2nd edition (2502)

ISBN: 9781718658851

Editor: Amy Hereford
Cover Design: © 2025
Page Design: © 2025

TABLE OF CONTENTS

Preface ix
   Dogtown Ecovillage ix
   Our Goals ix
   This Book x
Gardening 1
   Gardening Basics 1
      Native Plants 3
      Perennials 4
      Edibles 5
      Air, Earth, Fire and Water 7
      No-Dig 9
      Weeds 10
      Insect Pests 11
      Compost 15
      Sheet Mulch 17
      Start Small 19
      Companion Planting 20
      Pruning 21
      Fruit Flow 23
      Berries 24
      Herbs 26
      Butterfly Garden 28
      Establishing a Native Wildflower Garden 30
      Squirrels 31
      Harvest 32

- Fall Cleanup...33
- Permaculture...34
- Plant families..36
- Season Extension..37
- Watering...40
- Rain Water Harvesting..................................41
- Garden Markers..43
- Natural Building and Borders.......................43
- Hydroponics...46
- Plant Propagation..47
  - Seed Starting..50
  - Cuttings..51
  - Division..52
- Animal Life...53
  - Pollinators..53
  - Birds...54
  - Pond...54
  - Urine..56

Home...59
- Food..59
  - Organic...59
  - Plant-Rich Diet...61
  - Glass Jars...61
  - Sun Tea and Solar Cooking..........................62
  - Camp Coffee and Cold Brew........................62
  - Preservation: Solar Drying...........................63
  - Preservation: Fermentation..........................65
- Household..67
  - Reduce, Reuse, Recycle................................67
  - Cleaners..68

- Ants and other Household Pests..................70
- Cleaning Supplies..................72
- Basketry..................73
- Making Cordage from Foraged Fibers..................78
- Looped Bag from Foraged Fiber..................80
- Computers..................81
- Energy..................83
  - Home Energy Audit..................83
- Community..................85
  - Ecovillage..................86
  - Potluck..................86
  - Outreach..................88
  - Empowering..................89
  - Consensus..................90
  - Economics..................90
  - Co-ops..................93
  - Tool Sharing..................94
- Conclusion..................97
- Authors..................99
- Top Ten..................101
  - Top Ten Native Plants for Shade..................101
  - Top Ten Native Wildflowers for Sun..................101
  - Top Ten (Mostly) Native Fruits..................102
  - Top Ten Vegetables for Home Garden..................102
  - Top Ten Native Trees..................102
  - Top Ten Culinary Herbs to Grow..................103
  - Top Ten Medicinal Herbs to Grow..................103
  - Top Eight Animals for the Urban Garden..................103
  - Top Ten Plus Sustainability Tips for Renters..................103
  - Top Ten Home Energy Saving Tips..................104

Top Ten Garden Tools..105
Top Ten Ways to Reduce Global Warming..........105
Top Ten Backyard Birds in the Midwest..............105
Top Ten Backyard Butterflies in the Midwest......106
Top Ten Edible Weeds for Salads........................106
Top Ten Ways to Reduce......................................106
Top Ten Ways to Reuse..107
Top Ten Items to Recycle....................................107
Top Twelve Plants for Baskets and Fiber.............107
Top Ten Ways to Increase Garden Biodiversity...108

Preface

Dogtown Ecovillage

Dogtown Ecovillage is a community group in the Dogtown neighborhood that seeks to live sustainably, to learn and share knowledge and practices for sustainability. We are an urban ecovillage situated in a city neighborhood. We work together and help each other on sustainable living and group projects. We each own/rent our own homes and we gather for meals and projects, etc. The core group is all within a few blocks; others come from the surrounding neighborhoods.

We are located in the historic Dogtown neighborhood of St. Louis, MO, located just south of Forest Park and on the western boundary of the city of St. Louis.

Our Goals

1) Bring together and build a community of neighbors interested in environmental sustainability through shared experiences such as pot-luck dinners, community service projects, parties, and educational events

2) Enhance the environmental sustainability of the neighborhood and promote these values more regionally.

3) Create projects such as: gardening in private and shared settings, permaculture, home energy audits, improved housing efficiency, cottage industry such as aquaculture, environmental or social justice activism, tool lending library, shared car, shared wifi, shared child care, shared space for environmental activist organizations.

Our ecovillage is very loosely organized. More about that in the chapter on Community. After several years, we started to informally identify a *village council* of the more active members of the group. Often when an idea or project comes up, we'll run it by the council first. If it flies there, we might still tweak it before floating it in the wider ecovillage.

We don't have an official membership roster, dues or voting rights. We have an email group and a facebook page. If you're interested, you can join either of those and show up for events.

THIS BOOK

This book was conceived on Earth Day, 2018. Dogtown Ecovillage had a booth at the Earth Day Festival in Forest Park, St. Louis MO. This is one of the nation's largest and longest-lasting Earth Day celebrations.

Over the two days of the festival, ecovillagers took turns staffing our booth, sharing comaraderie and visiting other booths at the festival. In the booth we showcased several and group projects. DIY seed-starting pots, bee-hives, home-energy-audits, quail, vermicompost, an energy footprint quiz, coloring pages for the kids. We also fielded hundreds of questions on all of the above projects as well as questions about gardening, ecovillage life, composting, seed-saving, etc. We had lots of requests for more information, something in writing and other types of follow-up.

At that point, we realized that we have a lot of collective wisdom that we could share and that's how this book came about. It's written by ecovillagers and friends, and attempts to share our experience. Much of what we learned is from others, from books, from YouTube, from trial-and-error. We share it as the experience of trying to *live green* in our urban neighborhood ecovillage.

We are in St. Louis, so the book will focus on what works for us in Zone 6-7, in an urban setting, with our model of ecovillage. It's not so much the ultimate book on sustainability, as it is a snapshot of what we have learned about living sustainably both individually and collectively.

The book has four chapters: Gardening, Food, Energy and Community. In each chapter you'll find a range of topics, guided by questions we have received over the years, information we've come across that seems that we have lived, digested and simplified, innovations that become a part of every persons life when they decide to live greener. We're passing on what we know in hopes that we can all be better stewards of this great green planet we share.

GARDENING

GARDENING BASICS

Gardening is an important part of living green. Many ecovillagers work to make the green space around their homes, well, *green*. We've discovered that we don't have to abandon our cities and live in rural areas in order to be more sustainable. In fact, by living more sustainably right here in our urban neighborhoods, we can help our neighbors to make incremental changes as well. We can work to change policies and perceptions about various green practices. We can showcase and share what we've learned and increase the impact of what we're doing.

We have various themes and practices, but it's fair to say we generally focus on plants that are native to our bioregion, plants that are perennial, plants that are edible, and those that produce fiber. Actually, the best plant would be one that is native, perennial and edible. There are lots of examples: blackberries, rasberries, currants, pawpaws, sunchokes, and the list goes on.

The normal urban and suburban lawn that we grew up with is mainly turf with a few bushes, flowers and trees. The lawn is mowed every few weeks, the leaves are raked in the fall and we weed and tidy up our flower beds on occasion. Often the few species that are favored for these landscapes are plants that wouldn't tempt a single bug to munch on the leaves, and rarely draw a pollinator. And if any critter dares to intrude, they are often met with a barrage of toxic chemicals. Little by little this takes the life out of the system and out of the soil, and since insects are needed for birds, they don't visit as much either. Excluding these visitors also excludes the fertility that they add to the ecosystem, so we have to bring in chemical fertilizers as well. And this system also requires regular watering. I call this a chem-lawn. The average home may have only ten plant species.

Enter the ecovillager – we're looking for as much life as we can support. We gradually replace lawns with beds of native flowers and beds of veggies. We add native shrubs and trees. We mulch and compost to build soil. The result is a rich diversity of plant and animal life. One small urban lot may have one to two hundred species of plants and a similar diversity of insects and birds. It rarely needs chemical pesticides and fertilizers because it maintains itself naturally like a forest or prairie. Gardners intervene to tweak the system. Adding a plant here, mulch there, and pulling the occasional weed. Because of the insect diversity, garden pests rarely get out of control, and we learn to accept a little damage here and there.

In the sections that follow, we'll treat various topics and questions that come up as we work on "greening" our greenspace.

NATIVE PLANTS

Native plants are those that occurred naturally in a particular region, before the arrival of Europeans introducing plants from other parts of the world.

Native plants are particularly adapted to the climate, soil and insect populations of this region. They help to support our native birds, bees and butterflies who evolved in a dynamic relationship with these plants. Native plants have deep roots that enable them to draw water and nutrients from deep underground. They also support a diverse soil ecosystem, which helps to feed plants and increase the underground biodiversity. This creates a living soil that absorbs rainwater better, decreasing the run-off from storm events. It also filters and cleans the water which may eventually seep deep underground to replenish aquifers.

Native plants take time to establish. The saying goes: the first year they sleep, the second year they creep, the third year they leap. Some native wildflowers will only grow a few inches their first year and won't flower till their second or third year. You can get around

this by starting small (see section on Starting Small) and by buying two- or three-year-old plants.

But once established, native plants are easier to maintain. Select the right plant for a particular spot and you can enjoy it for years to come with minimal maintenance. This can cut down on lawn-mowing and hedge-trimming, which require inputs of energy. And it can also cut down on water needed to keep lawns green through our hot, dry summers.

Perennials

Perennials are plants that live for many years. Other plants, called annuals live their entire life in one year and produce seeds that are planted the following year. Perennials have several advantages over annual plants.

- First of all, you can plant it once and enjoy its beauty and its produce for many years.
- Because you don't have to plant over and over, you don't have to repeatedly dig and disturb the soil. (See No-Dig)
- Perennials build soil life and provide habitat for birds and pollinators, and other critters that bring life to your landscape.
- Perennials have down-sides: there is more work in the first year, and it may take a few years before the plant flowers or produces a harvest.
- After that first investment, you can enjoy the harvest for many years with less work to maintain the plant.
- Many perennials have deep root systems that are able to draw up water and nutrients from deep in the soil.

- Perennials may be fruit trees, native trees, shrubs, wildflowers or grasses. Some of these have root systems that are very deep, extending deeper each year. This helps to sequester carbon into the soil, thus doing something to help combat global climate change.

- Some perennials are the first harvest of spring, e.g. asparagus, rhubarb, sorrel and loveage. While other plants are just sprouting or struggling to grow enough to take a first cutting – the perennials have already provided an abundance.

- Beginning gardeners might want to start with annual plants that give a harvest the first season. It's also a great idea to add a few perennials each year. Little by little you can increase perennials and when they start to yield, you can decrease some of your annual veggies.

Edibles

Edible plants are the third part of the sustainable gardening triad. Food grown in our back yard travels several feet from "field to fork". Whereas most food in a grocery store has traveled thousands of miles. Even food in our farmers markets may have traveled a hundred miles or more. Each mile the food travels adds to the carbon footprint of our food. No carbon is expended in bringing a tomato from your yard to your kitchen. The same is not true of the tomatoes in a local supermarket.

Recently, there was a massive recall of lettuce. Not to worry. Though it was early spring, we were harvesting our greens from the yard – worry free. Veggies were not only free of E. coli, they were free of pesticides and were nutrient rich and organically grown.

Speaking of organic, many of us in the ecovillage choose to grow organically. There are several reasons – 1) chemicals we use to kill one bug or plant, or fertilize our crops, also kill other bugs and plants, harming pollinators and song birds, sometimes remaining in the ecosystem for years. 2) these chemicals also harm the soil-food-web that provides nutrients for our plants, thus we would require more chemicals to feed the plants. 3) food grown organically is more nutrient dense – it packs more micro-nutrients than conventionally grown food – it's better for you, according to growing research by dietitians who study the link between nutrition and disease.

Another reason for growing edibles is that you can grow what you like and you can grow as much as you like, within reason.

Food is also super fresh – go out to the garden to figure out what you are going to prepare for the next meal. Pick it, clean it, prepare it and eat it. Food that fresh has a richness and flavor that you just can't get from food that is harvested weeks earlier, before it was ripe, and shipped for thousands of miles.

> *"As a kid, each year my Mom would say: we never have enough of a particular vegetable. One year is was peas, then cucumbers, then something else. I always took that a challenge to make sure we had enough of whatever it was.*

Growing your own also means less waste – just pick what you need – and share what is extra. A bumper crop of strawberries can make for good neighbors. Growing your own also means less packaging. You will cut back on waste and recycling by growing more of your own food.

One strategy is to grow enough to eat, but not much to preserve. In the off season, you can go back to buying produce. If you have an abundance, see the later section on Preservation.

AIR, EARTH, FIRE AND WATER

Some of us have green thumbs, probably a lot of us in the ecovillage, but not all. Some claim to have brown thumbs. This is something I have a hard time understanding. Plants know what to do – they just grow. If we stay out of their way, they will pretty much do what they do – and grow! Probably the brown thumb club is a little too caring, more plants have probably succumbed to too much water and too much care – than too little. You also have to put a plant in a location where it can have the best chance of survival. Some like sun, others like shade. Some like wet spots, others like dry areas. Putting a plant in the right spot is the first step in having it thrive. When you get a new plant, check where it was growing before, or check the label that came with it. Look online to see recommendations for that plant for your area. Then go find a place in your garden that meets the plant's needs and plant it there. Well begun is half done!

The second step in gardening is to give your plant the right amount of four key elements: air, earth, fire and water.

- Air – a plant needs moving air, it needs to not be crowded too much by other plants. And it needs to have an open branching structure (See Pruning).
- Earth – not surprisingly, a plant needs soil. Make sure that there is enough soil for the plant to dig its roots in, and make sure there is some organic matter in your soil. A soil test is not a bad idea either, to enable you to add the right building blocks in your soil.

- Fire – plants need energy in the form of sunshine. They need the right amount of light, and the right temperatures. Some plants like peas, spinach and lettuce, thrive in the cool temperatures of spring, but wilt and die in the dog-days of summer. Other plants, like sweet potatoes and black-eyed peas grow like nobody's business when it's hot and sticky, but won't grow and inch in a crisp fall day. Here again – check out what the plant wants, and plant it where and when it will thrive.

- Water – all plants need water to survive. Some plants like to sit in water, others like to get good and dry in between waterings. Know your plants' needs and provide accordingly. I site my perennials and natives so that they get what they need without extra water, except in extreme drought. But I do supply water to my annual vegetables, on a timer, with a drip line. More about that in the section about Watering.

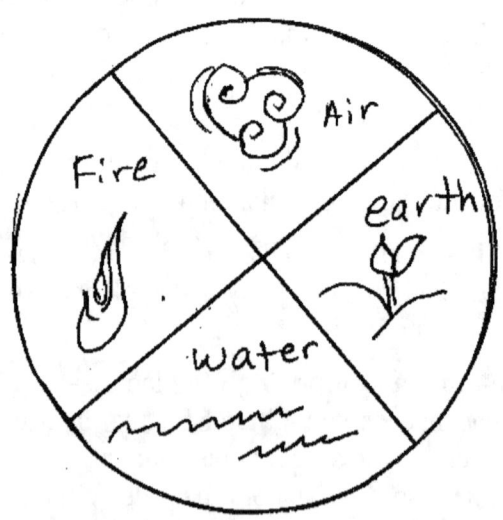

No-Dig

Traditional gardening required the soil to be turned or cultivated each season before planting could begin. The idea was to break up the soil, aireate it and mix in compost or other amendments. This is effective in some regions, but is less so in hot dry climates. Some gardeners have moved to a style of gardening called no-dig – meaning that you disturb the soil as little as possible. This helps to preserve and build the soil structure.

In my area, we have heavy clay soils with basic soil particles that are quite small. Soil structure refers to the way these tiny soil particles are bound together by physical, chemical and organic means. When you disturb the soil by digging, you damage this soil structure. Those of us who decide to use a no-dig gardening technique do so to preserve the soil structure and microbiology. If you think of the way soils developed and have been preserved in nature, they are rarely dug. There may be major events like a tree being uprooted by wind – that will change soil structure. Also, small mammals and birds will dig and burrow. Worms and microorganisms in the soil will also move soil particles. However, there isn't a consistent breaking up of a whole bed of soil in nature. And natural soils usually have much better structure than cultivated soils. There was a time when we thought that cultivation areated the soil and made it easier for plant roots to penetrate. However, many people now question the wisdom of this practice. Some trials show better garden performance in no-dig beds.

In addition to maintaining soil structure, no-dig also avoids bringing weed seeds to the surface, so it is generally less weedy. Mulch and compost are added to the surface and allowed to gradually work their way down through the action of rain seeping down and critters who accomplish small-scale churning of the soil.

The final reason for no-dig is that it is less labor intensive. You don't have to till or dig the soil each season. Just add mulch to the top and allow nature to pull it down.

Weeds

"A weed is a plant whose virtues have not yet been discovered" says Ralph Waldo Emerson. Weeding is probably the most maligned task of gardening. Every gardener has pulled their share of weeds over the course of a lifetime. A better strategy is to manage the garden in such a way that there aren't many weeds. Then it's just a matter of pulling the odd weed as you visit the garden to do other tasks.

- A first line of defense is mulch, and particularly sheet-mulch (see Mulch). Mulching with wood chips makes it harder for weed seeds to germinate. Put enough mulch on top of the weeds, and the weeds then become compost too.

- Gradually learn to identify weeds. Never pull something you don't recognize. There are a small handful of early spring weeds: henbit, dead nettle, chickweed, and speedwell. They don't last, so you can just try to thin them out. Other weeds are hardy perennials. Eliminate them once, and they are gone for good.

- You might have a few zero tolerance weeds – e.g. some that are particularly aggressive or those that you are allergic to. You will want to pull them any time you see them. Others are native plants, and you can be more tolerant. When you begin cultivating a new space, choose your battles. Eliminate one type of weed, and you may only rarely see that one again.

- If there is 'weed pressure' from neighboring yards and garden, then make a strong border to keep weeds from making their way back in.
- Some weeds are our friends. They can be great edibles: wild violets, purslane, lambs quarters. You can let a few of these flourish to provide a trouble-free source of greens. Make sure you have some plantain on the property as well. It is a medicinal that soothes cuts and stings.
- Finally, make your peace with some weeds. For the most part, you can manage them out of your system, but if there are a few that stay, so be it. If they are on your zero-tolerance list, then they go, otherwise, let them have a little space.

Insect Pests

Well now, you can't really talk about gardening without talking about bugs and the many ways of dealing with them in your garden spaces.

There is an approach to insect pests called *integrated pest management (IPM)*. Under this system, gardeners, farmers, orchardists and plant-people in general take a multi-prong approach to dealing with pests. They try to avoid using chemicals, but in extreme situations, they may use them in a very limited and targeted way. That would generally be favored in the ecovillage, and we probably expand on it a little. Our approach is, biological, physical, cultural and chemical, and we would probably add psychological. Let's start there.

I take a fairly live-and-let-live approach to bugs. It's true, there are a few that you may want to work against, but for the most part, you could come to accept them as a part of nature and the ecosystem. Take an interest in bugs, get to know them, know which critters are beneficial and which are more harmful. Get to know when they show up and what plants they eat. Get to know who eats them and how to attract these predatory insects. The world of bugs is facinating and curiosity is a great place to start in the psychological 'war on bugs.' If you're planting native plants, you have hundreds of species in your landscape. That's the point of native plants. So bring on the critters. And having a wide diversity of critters around usually helps limit the harmful pests. Another psychological strategy is to learn to live with some pest damage. Plan to grow enough to satisfy yourself and the pests.

More conventional IPM starts with biological methods. This ties into the psychological methods, and seeks to invite a wide range of critters into your gardens, by planting a range of native plants (See Native Plants, and Start Small). As you open your system to more bio-diversity, you'll probably find that the harmful pests come down to a manageable level. You can also plant fragrant plants (e.g. herbs and marigolds) to discourage loitering insects. Another strategy is to intermix your plants, so that a pest on one plant can't just move plague-like down your neat veggie rows. Another option is to let the chickens into your garden, but just long enough to eat the insects, not long enough to decimate the plants.

Physical control seeks to actually physically remove the insects from the plants. You can pull larger bugs like tomato hornworms off plants, particularly when they are decimating the plants. It's also possible to blast pests off with a burst of water, or sometimes you can brush them off with your finger or shake them off.

Cultural controls are a range of good gardening practices that limit the presence and spread of disease and disease-bearing critters. They range from crop rotation, to removing specific invested or diseased plants, to using mulch to avoid water splashing onto plants, to reducing standing water that could breed pests.

Crop rotation is a big idea, but it can be fairly simple to implement. Some plants are perennial and they will stay where they are planted. However, most veggies are annual plants that can be moved from place to place in the garden from one year to the next. You have to work with the space you have, it would be great if you had 10 beds and 10 crops and you could rotate crops through the bed so it was 10 years before any bed had a repeat planting. However, many of us don't have that luxury. Also, not all plants are equally suited to all of your locations. Nevertheless, it is helpful to establish a few different areas in your garden, three to five areas will do.

- The first area can be for spring greens, turnips, radishes and carrots. Some of those will complete their growth in four to six weeks, others will take a few months before they are harvested. Let them grow then mulch the area over and repeat with a fall planting of the same.
- The second area or bed can be for beans, tomatos, cucumbers. The larger summer vining plants that are often staked for better growth and harvesting.
- The third area can be used for melons, squash and sweet potatoes that tend to sprawl across the ground.

This three bed system can be expanded if you have enough space to divide some of the beds in two parts – then you can rotate the beds and switch up the crops from side to side in a bed.

So we've covered psychological, biological, physical and cultural methods. Now for a word on chemicals. First there are the home-remedies. Spraying soapy water on plants will often stop chewing insects, a sugar-borax-water paste is an effective ant trap, beer can attract slugs, you can use a smelly tomato stem to brush aphids off other plants. These are all natural remedies that kill a few bugs, but don't degrade the environment. Insecticides are another matter. They often result in a 'scorched-earth' situation, killing all insects, even butterflies, bees and other pollinators, killing birds and mammals who eat insects, killing soil microbiology that feeds our plants, and harming the unsuspecting child or pet unfortunate enough to ingest sufficient quantities. Let's think of this as a limited-use, last-resort option. Start at the top of the IPM list, by the time you get to insecticides, all but the most pesty of the pests will have given up, moved on, or completed their lifecycle. Sometimes I decide not to grow something, because the critters get it before I do. Try something else – there are lots of options that provide enough abundance with less opposition.

Compost

Compost is decomposed organic matter that is used to improve soil quality, to fertilize plants and to help prevent garden pests. Garden centers sell compost and many cities compost yard waste and make it available to area residents, often for free.

Compost can be added to the top of the soil, and it will gradually be worked in through mechanical and biological processes. You can also add a little compost to the hole each time you plant something into the garden. Often there is compost on the top of the soil, so make sure some falls in as you dig a hole for a transplant.

Composting is also a process that we can use to divert our own organic waste from landfills and turn it into healthy additions to our soil. Many garden writers have written extensively on how to compost. However, it seems that too many people then find it so complicated that they never get started with it. Composting is as simple as leaves falling to the ground in the woods. No one rakes them into a pile or makes sure the proper balance is there. But those leaves break down – simple! So whatever you do, make sure that you start composting. Let's look at a few principles, then some simple possibilities.

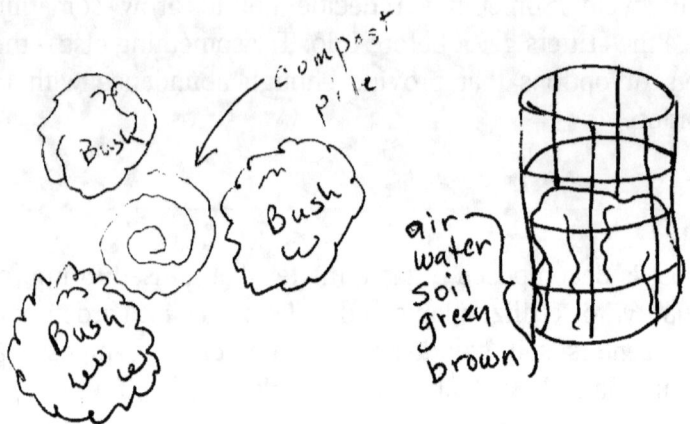

First you need a few basic things for compost:

- Soil – make sure your compost is touching the soil – that will provide the microorganisms needed to get the process going. Or toss some soil in your bin.

- Fresh plant matter – also called 'greens' or nitrogen. For best results, this should be un-cooked plant matter. Avoid oils, meat and dairy, though small amounts don't usually cause a problem, e.g. less than 5%.

- Old plant matter – also called 'browns' or carbon. This can be leaves, paper, cardboard, egg cartons, etc. Don't add wax or shiny paper.

- Water – the pile should be moist, but not soggy. So in damp weather you may need a cover, in dry weather, you may need to water.

- Air – the microorganisms that do the composting need some air. If your pile is a yard in all directions, you may need to occasionally turn it; if its smaller, you won't need to.

You can make compost faster if you have at least a cubic yard and you actively monitor temperature and moisture and turn it on a regular schedule.

However, most of us don't have enough to make active management an issue. Put up a ring of fence wire that is about two feed across and three feet tall. This is set on bare soil and toss compost in there. Leave it uncovered. Have a pile of leaves (browns) nearby so you can toss a few on top after you dump your kitchen waste (greens).

You can use the same basic procedure in between a cluster of three or four bushes. Just toss the compost between the bushes without the fence ring. It works like a dream. At the end of the year, you might dig the compost in a bit, or just cover it over with leaves. You don't have to cart it around to other plants – but you might bring it to a plant that needs a little extra nourishment. Put a small pile near the plant then put a brick or stone on top of that. In a few weeks' time, the compost will have disappeared into the soil.

Sheet Mulch

Mulch is a process of placing organic matter on top of the soil to moderate soil temperature and moisture for the benefit of plants, and to suppress weeds. The mulch gradually breaks down and enriches the soil as well.

Sheet mulching is a modification of this. Prior to adding the layer of organic matter, lay down several layers of newspaper or a layer of cardboard. This further aids in suppressing weeds by blocking light.

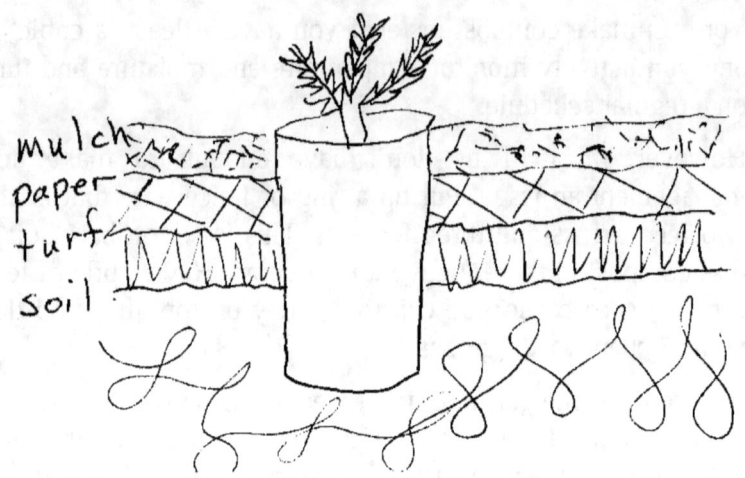

There are many ways of doing this. Try placing cardboard over turf, and placing an inch or so of mulch on top of that. This will effectively kill the turf, which then composts in place. To plant, dig through the mulch, poke a hole in the paper or cardboard and plant into the soil beneath.

In succeeding years, an additional layer of paper or cardboard can be applied, and more mulch added to the top. In the second and succeeding years, the layers can be thinner. And it may be possible to simply add additional mulch without the paper or cardboard. This will depend on how much "weed pressure" you have in a particular area. You can paper and mulch around the edges, where the beds meet the lawn, and only lightly mulch as needed further into the bed.

After several years of this process, the soil structure will improve as the organic matter breaks down and is increasingly incorporated into the soil. At the same time, the number of weeds will decrease since the seeds below gradually die away, and seeds landing on top of mulch do not usually find good conditions to germinate.

Start Small

Once you get the sustainability bug, you may want to try to convert your whole yard to natives, perennials or edibles. That's a great goal, but it's usually a better idea to start small. Even a single garden bed can make a difference. And by starting small, you're guaranteed that all your starting mistakes will be small as well.

For you, small might mean a 20 square foot butterfly garden, or 10 square foot rain garden, or a few tomato plants, or a berry bush. By starting small, you can plan your first move wisely. What do you have time for? What do you have room for? Do you have sun or shade, wet or dry? Would you rather start with native plants, or with edibles? The answer to all these questions will direct you to your first small step into a more sustainable landscape.

Starting small also means you'll have the time to learn how to care for your new plants. You'll have time to learn what works best for your goals and is with in your budget for time and money. All this means your more likely to be successful and you'll be more likely to continue the project next year, and maybe even expand it.

Starting small doesn't mean you don't want to have some long term goals. If you would like to do five or six things right away, maybe it's best to break that down into a few smaller projects. Start one project in the spring, and plan to start another in the fall, and maybe a few more next year.

Projects often take more time the first year. Preparing the site, selecting the plant, installing it and babying it through the first few weeks till it gets established. Then the following year, that project may just need some maintenance, and you can take on another project.

Companion Planting

Gardeners generally have a list of what goes with what companion plants assist in the growth of others by attracting beneficial insects, by repelling insect pests, or providing plants with nutrients, shade, or support.

One example of this is the "three sisters" garden. In this technique, gardners plant corn, that provides support and needs nitrogen, with beans, which need support and fixes nitrogen, with squash which sprawls under the corn and beans, helping to suppress weeds, shades the soil, increasing soil moisture, and discourages some pests from attacking the corn and beans. These three plants are "sisters" that have each others' backs in the garden. They also provide a rich diet of carbohydrates from corn, protein from beans and vitamins from squash.

There are lists of companion plants, planting wildflowers in the garden helps to increase pollinators and other beneficial insects. Planting herbs with their strong smells can repel chewing insects.

There are also lists of antagonistic plants – plants that attract the wrong insects or compete for nutrients or sunlight or space.

A simple way to do this is to group veggies that you put together, then rotate that group, if you are doing "crop rotation." (See Insect Pests above.)

You can grow peas up any easily accessible fence since these are planted in early spring. They can go in any bed, just keep them away from the onions.

This simple rotation keeps companion plants together, and antagonistic plants apart. Then you can plant flowers and herbs throughout the space. If you want more detailed information, there are plenty of lists of companion plants on the internet.

Pruning

Good pruning is both art and science. You prune a tree or shrub to maintain and promote its health, to increase blooms or fruit, to maintain its size and balance.

Begin visual inspection at the top of the tree and work downward. Timing of pruning depends on the plant species, and your goals, but generally it is best to prune in late winter when plants are dormant. It's a good idea to search books or the internet for guidelines on pruning the specific plant. Here are some general guidelines:

- Ordinarily you never remove more than ¼ of a tree's crown in a season.
- Ensure the main side branches are no more than two-thirds the diameter of the trunk.
- Remove defective parts before pruning for form.
- Where possible, try to encourage side branches that form angles at ten-o'clock or two-o'clock with the trunk.
- For most species, the tree should have a single trunk or central leader. Identify the best leader and lateral branches before you begin pruning.
- Don't worry about protecting pruning cuts, the tree will heal itself. For aesthetics, you may feel better painting large wounds but it doesn't prevent or reduce decay.
- Keep tools sharp. Hand held pruning shears with curved blades work best on young trees.
- For high branches use a pole pruner. A major job on a big tree should be done by a professional arborist.

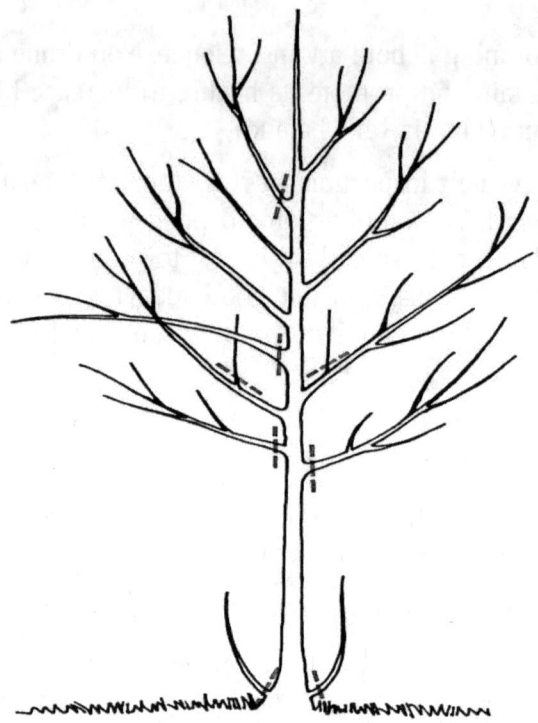

- For larger branches, cut outside the branch bark and ridge collar (swollen area). Do not leave a protruding stub. If the limb is too small to have formed a collar cut close.

- When shortening a small branch, make the cut at a bud or branch. Favor a bud or branch that will produce a branch that will grow in the desired direction (usually outward). The cut should be sharp and clean and made at a slight angle about ¼ inch beyond the bud.

- Try to find an opportunity to shadow a professional arborist, trying to understand the overall strategy, and why and how each cut is made.

- Learn as you go. Your trees and shrubs will be with you for years, so keep an eye on how your tree grows each year after pruning. Practice makes perfect.

Fruit Flow

The first lettuce of the spring garden is a special treat, as is the first strawberry to ripen. From the moment of these early harvests, plan your garden so that you will have a steady supply of fruits and veggies throughout the spring and summer and into the fall. Perennial veggies help you get a good early start with asparagus, loveage and sorrel. I go for steady production, rather than bumper crops that have to be preserved.

In planning your garden, look for early, mid and late harvests. Plant a week or so before the weather is right – just in case you get lucky and get an early harvest. Then plant a second and third planting of lettuce, peas, spinach, and all your spring veggies. This should ensure that you have a steady supply from the backyard, and you may not have to buy produce from the market.

Look for the same steady supply of fruit from the garden. Think of this as *fruit-flow*. Strawberries are the first to ripen, followed by early and late blueberries. Next come the blackberries and raspberries, and so on through the season.

You can have smoothies from "backyard berries" for several months. Dry and freeze some (see below), so that you can extend the smoothie season.

This takes planning:

- Plant various fruits that ripen throughout the season.

- Select varieties that ripen early or late to complement other fruits that are ripening.

- Place the same variety in two different areas of the garden, so that they ripen at two different times. Plant near a brick wall, and the plant is likely to bloom and fruit earlier than the same variety in an open area.

This can't be planned perfectly, and the weather will afffect different plants differently, but I work toward enjoying a steady stream of fruit flowing onto my table throughout the summer.

Berries

Often a gardener's easiest foray into growing fruit will be berries. While fruit trees seem to be a major commitment in terms of time and money, a few berry plants are less of a commitment. In fact, most of the fruits that are native to North America are berries: blackberries and raspberries, blueberries and cranberries, currants, gooseberries, elderberries, mulberry and many, many more. Being natives, these plants are well adapted to various regions of the country, and many can be added to an urban lot quite easily. They will generally bear fruit after only a year or two, while it may take 4-5 years for a larger fruit tree to bear.

Many gardeners will plant strawberries as their first attempt at fruit. There are native strawberries, some of which are more flavorful than others. However, the common garden strawberry is a cross between a North American strawberry and a South American strawberry – and the cross was done in France; it's quite a cosmopolitan berry. June-bearing plants are more robust plants with larger yields, even if ever-bearing varieties sound tempting. Obtain plants – often provided as bare-root plants. Set them in the soil as soon as you obtain them and water them in well. Pinch back the blossoms the first year, in order to give the plants an opportunity to build vigor. Then the second year, you can enjoy a harvest. In our climate, we mulch strawberries with a layer of leaves in the winter, then pull back some of the leaves in the spring. Leave some leaves to mulch the bed to enrich the soil, retain moisture and suppress weeds. Beds are best renewed every 4-5 years to ensure continued production. You can also extend your berry season by varying the micro-climate of your strawberry bed. Some gardeners put stones in the sunniest part of the bed, to encourage early blooming and fruiting. Another part of the bed could be shaded, or perhaps the mulch could be left on longer so the bed will fruit later and continue fruiting when the warmer section has finished giving berries. Alternatively, early and late varieties could be used.

Other berries grow on bushes or brambles. Like strawberries, they should be given one growing season to settle in, and may begin bearing fruit in the second or third season. Some gardeners use bird netting to protect their berries, but others are willing to share the harvest with birds, so long as there is enough for the family. Most berries benefit from pruning in the second or third year. It is good to check with each species to determine the best method. Some berries fruit on first year branches, others on second or third year branches, but older branches become barren. You'll want to learn how to prune the particular plants you have to maintain the health of the plant, increase yield and manage the size of the bush. As with strawberries, mulch is an important part of retaining water and suppressing weeds.

Herbs

If you only have a small space, then herbs are a good choice. You can grow some great herbs on a balcony, or even a sunny window. There is no comparison between fresh and dried herbs.

Most herbs are mediterranean plants, so think of a sunny, well-drained location. Some will tolerate more or less water, more or less sun. Many herbs are perennial, though there are a few great annuals as well.

Chives, thyme and oregano are great perennials and are super easy to grow – in fact you may have more trouble containing them. Mints are great, but make sure you're ready for their sprawling nature. Lavender is a wonderful herb, but can be picky about growing conditions in my region.

Basil and cilantro are equally good annuals. They will generally re-seed, i.e. they will go to seed and the seeds will drop and sprout the following year. Again, you may have a challenge containing them.

There are two herbs that I keep trying, though I don't have great luck with them: rosemary and bay-laurel. Neither is cold hardy, so they do have to be brough indoors in our zone 6-7 winters. However, rosemary has an stunning fragrance, and fresh bay leaves are amazing – so much better than dried bay leaves.

Herbs tend to have few pests because of their strong scent and taste. The very things that make them great for cooking also make them distasteful to other critters. Pollinators generally love them, because they flower profusely through the heat of the summer. Because of their tendency to repel pests, you can plant them by your veggies to help control insects.

Cooking with fresh herbs can be delightful. Start growing an herb you love in your food. Fresh chives with butter on a potato is amazing. Fresh basil with a little oil and vinegar is great with your home grown tomatos. You don't have to start with every herb, just start with one you love – chives and basil are great choices, as are oregano and thyme.

Once you have a few fresh herbs in your garden, try adding them in combinations of three to five herbs. Take a bundle of the herbs and wrap a string around it and pop it in your soup or stew when it's cooking. Pull it out before serving, like you would a bay leaf. Experiment with amount and mixes of herbs.

Herbs are also medicinal, and some herbs are great to keep in the home garden. Here is my top ten list:

- Chamomile is a soothing tea made from the plant's delicate white flowers that can be grown in a sunny dry place.
- Feverfew is good for digestive problems and headaches.
- Lavender smells lovely, is a pollinator magnet and has a calming effect.
- Lemon Balm has a wonderfully lemony fragrance, is good for teas and is good for anxiety and insomnia.
- Mints also make great teas and are relaxing and may provide relief from headaches.
- Thyme is a strong anti-michrobial – use it on cuts and wounds.

- Plantain is good for skin conditions. Often considered a weed – I tell kids its a 'bandaid-plant.' Keep a few around the garden – they are great for pollinators and when you get a cut, sunburn or insect bite, crush a few leaves and put it over the area for immediate relief and to promote healing.
- Passionflower leaves make a tea that can be used for insominia, headaches and other types of pain.
- Bee Balm can be used to make a tea or can be infused in alcohol or vinegar. It is anti-microbial and anti-inflamatory.

There are lots of other herbs that are easy to cultivate in the home garden, as well as many books and websites that offer advice on how to use and how to grow them.

Butterfly Garden

A butterfly garden can be as simple as planting a few flowering plants and can be as complex as planning a complete habitat that can be certified through several national conservation programs.

Native wildflowers that bloom throughout the season will provide nourishment for adult butterflies. It is important to choose native plants and to ensure that your plants were grown without the use of pesticides. Knowledgeable garden experts can help you find sources of native plants for your area.

You will also want to plant caterpillar food plants. While butterflies will often take nectar from a variety of plants, they are very picky when laying eggs. They will generally lay their eggs on a narrow range of plants species, to ensure the caterpillars will have plenty to eat. Yes, they will be eating those plants you're caring for. Most caterpillars will spend their entire caterpillar life on that same plant species. Then when the time is right, they will wander off to some other place to enter a chrysallis where they will be transformed into a butterfly.

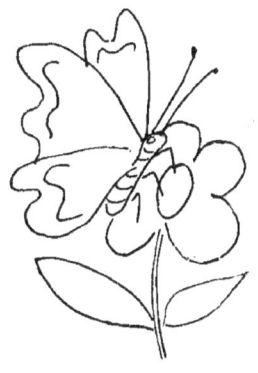

The Xerces Society (xerces.org) and the North American Butterfly Association (naba.org) will help you know which plants are favored by which butterflies.

The Monarch butterfly is an example of this specialization. They will only lay their eggs on milkweeds. And even among milkweeds, they have their preferred species. If you're looking for monarchs, you'll want milkweeds in your plant mix.

If you have space for ten plants, try two or three plants of three or four species that flower at different times. If you're expanding next year, maybe you can add additional species. You may also find that some of your plants do better than others. It may be that some are slower to establish. It may also be that your conditions aren't right for some species you planted.

In one recent project, we planted several each of two spring bloomers: Phlox and Packera, two summer bloomers, Milkweed and Coneflower, and two fall bloomers: Aster and Goldenrod. In this particular installation, it was important that the plants look good, so we chose more compact plants. In any installation, I always try for asters and goldenrods since these two plants support a wide variety of pollinators.

In addition to the plants, you'll want to provide shelter, sun and water. Shelter can take the form of trees or bushes near your site. There should be some place for the butterflies to sun themselves to warm up for flight and for feeding. And finally a water source is helpful a sandy puddle is ideal, but make sure you're not breeding mosquitos.

There are plenty of online sites that will help you design a butterfly garden.

Establishing a Native Wildflower Garden

Establishing a successful native planting from seed involves a labor of love and patience. Most who have been through it will praise the process, the thrill of discovery and the absolute joy in transforming a space with little biological activity into a healthy eco-system. You will delight in your efforts season after season as the wildlife and color reminiscent of the North American prairie become a part of your home landscape.

Year 1: Site Preparation and Seeding: Many areas will need 1 growing season (spring-fall) for site preparation, an exception may be a garden plot, or new construction area. Remove existing grass/weeds and the weed seed bank that may be in the soil during the growing season (April-September) by smothering, repeated shallow tiling, using herbicides or other methods you determine to be best. Sowing the seed: I like fall or frost plantings (mid-Oct to mid-March). Spring plantings (April-June) are an acceptable second choice. You should not plant in the summer.

Year 2: First Growing Season: Most sites need maintenance mowings to keep weeds from going to seed and to allow light to penetrate the ground encouraging growth of the majority of the slow-growing natives. You may get some blooms this year, most likely annuals like Black-eyed Susan or Partridge Pea, but you must sacrifice these native flowers if you want the other species to establish. Keep the area cut to 4-6" this year. Don't pull the weeds! This will disturb the root systems of the nearby natives trying to establish.

Year 3: Second Growing Season: This may or may not be the year you start enjoying the fruits of your labor. You may need to mow once yet this year. Be patient, although some species may reach flowering stages in year 3, some could take 4 years or more.

Squirrels

Squirrels are a fact of life in our urban setting. And they can be a nuisance in our yards and gardens. They dig up plants, they eat our tomatoes and our fruit, and they gnaw on our stuff. So what can we do about this.

- One approach is the glass-half-full approach. When a squirrel is digging in my no-dig garden, he is doing the tilling that I have chosen not to do. So there is some mixing of soil for free.
- Anytime I plant something out, whether seeds or seedlings, I protect it from squirrels. I have several pieces of chicken-wire, ranging from 2'x2' to 2'x4'. Anytime I plant something, I lay the chicken-wire on the ground over the planting, whether seeds or seedlings. Then I use metal landscaping staples (2" metal staples) to hold it down. It stays in place till I need it for another planting.
- Another squirrel protection device is a wire basket. I have several of these that range from 6"-12" across. I place this over the seedling and weigh it down with a rock or hold it down with sticks or with landscaping

staples. I pick them up at yard-sales or second-hand stores. They move around the yard every time I put out a new plant.

- Yet another variation on the theme of squirrel defense is placing a few rocks around a newly planted seedling. Three or four rocks that are too big for a squirrel to dig can be placed around the plant until it is ready to hold its own. Again these can be moved after a few weeks to another plant in need of support. These rocks can also have the added benefit of warming the soil and holding in moisture. Both can be helpful when you are planting out seedlings in the spring.

- Some area gardeners have dogs, and they generally keep the squirrel population at bay, but don't really eliminate it.

- I haven't managed to stop squirrels from eating fruit and tomatoes. I have come to accept it as a price of having fresh, local, organic produce throughout the season.

- Some completely enclose their plants, but it is a lot of work. Make a 10ftx10ftx10ft cube frame out of conduit, and enclose the entire structure with chicken wire. Add a door frame and a chicken-wire door. It's a lot of work, but it might just do the trick.

Harvest

Like many garden and household tasks, harvesting takes time, but many consider it the best part of gardening, and its greatest reward. As always a few tips can be helpful.

- Harvest just before the meal if possible. The food is so much better when it is fresh from the garden.

- Harvest just what you need. If you are preparing a meal, bring in what you need for that meal. If there is more there, you can either decide to harvest and preserve or just let it go to the critters.
- This follows on the above – plant what you plant to eat, or perserve. There is no sense in going through the effort to plant, cultivate and harvest, if things will just end up going bad in the fridge. See the section on food preservation.
- I tell kids that my veggies are cleaner than water. Well, that may be a stretch, but it may also be true. A plump, ripe tomato, hanging on the vine after a rain storm is probably cleaner than my tap water. We wash veggies from the store or farmers market because we don't know what's on it – or many be cause we know there are pesticides on it. But think twice before washing your garden veggies.
- That being said, a wire basket is a great tool for cleaning root vegetables or leafy greens. Place the produce in the basket and then sink the basket in a bucket of water, right in the garden. It may take a few washes, but it is usually quicker and easier than bringing it to the kitchen as is. The water can be dumped in the garden as well.

Fall Cleanup

As fall comes on, the garden will slow down. Gardeners enjoy a few late season plants and produce, but by late fall, the garden is fairly brown and barren. It's time for fall clean-up. Gardeners get out in the crisp fall days and clear away the debris so that they will be ready for spring planting. Some plants need a little extra protection in order to make it through the winter. How do we do fall cleanup in a green way?

- First of all, you'll want to leave some plant debris for the critters. That may mean collecting leaves and piling them in garden beds and around trees and bushes. You can shred the leaves, so they are easier to work with and break down easier, or you can leave them whole.
- Leave the seed heads from your flowers, and you'll find birds visiting for a winter snack. If it looks too messy, you can cut the seed heads, bundle them, and place them around the garden.
- Leave some bare ground for ground-nesting bees and other pollinators. Keep a pile of old rotting logs and sticks where they serve as a nesting site for other critters.
- Fall can be a good time to renew your mulch and borders so you don't have to take time for these tasks in the spring.
- Clean and store tools, trellises, pots, etc.

Permaculture

Permaculture is a movement in sustainable gardening that embraces some fundamental orientations toward landscape, gardening and even society. It seeks to work with nature.

Observe the way natural ecosystems grow, without any human input. Many species grow together, sometimes competing for resources, sometimes sharing resources, sometimes exchanging resources.

It is not possible to explore all its practices and techniques in a few pages. However, it is possible to point to the many resources that are available on the topic. Some of these resources are thoughtful and well grounded in science. Others are not so well grounded, and are passed on from person to person without a deep understanding of science or experience on the land.

Some of the ideas of permaculture are found in other sections of this book, though they may not be linked to this movement. Let me list a few ideas that I have found useful:

- Observe and interact is one of the principles. In other words, don't just read the seed packet and follow its directions without seeing how the plant responds. By all means read the seed packet, and with all that in mind, observe how the plant responds to your garden, to your nurturing, and adjust accordingly. Notice areas of the garden that have more sun or shade, that are more wet or dry, that have more or less wind. The same plant may do better or worse, just by moving it a few feet to a spot that is more tailored to its specific needs.

- Keep resources on site. I have found that I can use long branches I pruned from one tree for edging, or for staking. Keep rainwater on site – don't let it run off into storm-drains without at least making an effort to slow it down and soak it in. Work with the sun's energy to store it through bio-solar panels – i.e. through plant leaves.

- Build soil, allow old plants and fruit to break down in a compost pile or bin. Tend to the soil, and the soil will grow the plants.

- Permaculture divides property into zones, based on how frequently you visit that space. Your home is zone 0, the space just outside the door is zone 1, next are garden beds in zone 2, fruit trees in zone 3, a semi-wild area in zone 4 and wilderness in zone 5. At first glance, this seems obvious. But when deciding where to place different elements in the landscape, ask how often you'll be going there. The herb garden goes in zone 1 just outside the door so you can pick herbs fresh when you are cooking. The fruit tree can be at a distance because it is largely unattended for much of the year.

Plant families

Plants families are groups of plants that share similar genetics as well as similar characteristics, growing conditions and propagation methods. There are vegetable plant families:

- Legumes: Beans, limas, greenbeans, peas, long beans and black-eyed peas. There are warm and cool season legumes, they all add nitrogen to the soil. Many benefit from some sort of trellis

- Alliums: Onions and garlic – usually planted from cloves or sets.

- Nightshades: Tomatos, peppers and eggplant and potatos – These warm season vines need an extra helping of nitrogen. The vines benefit from some support. Potatoes are cool season crops. Sweet potatos are a separate family and like sun and heat.

- Curcurbits: Melons, squash and cucumbers – warm season vines that need plenty of water. Cucumbers benefit from some support.
- Brassicas: Cabbage, broccoli, cauliflower, brussel sprouts, and arugula. These cool season crops can be started indoors and can be grown spring and fall.
- Lettuce, Spinach and Parsley/Carrots are three families, but it is convenient to treat them together.

Season Extension

As gardeners we are generally always looking for the earliest harvest in spring, and going for the latest harvest of fresh produce in the fall. As they get into their craft, they learn that they don't have to settle for a season that starts with the first frost and ends with the last frost. There are ways of extending the season by weeks or even months. This means more fresh fruits and veggies. One range of techniques is food preservation – discussed in a separate section. Here we'll talk about techniques to extend the growing season by growing earlier in the spring, and later into the fall. At the extreme, you could have fresh produce all year long, but in our climate that would mean adding light and heat – which brings up other sustainability issues. The section on Aquaponics will expore some of these options.

Season extending techniques vary from the simple to the elaborate, from the cheap to the costly. Here are some that we have used:

- Starting seedlings indoors in spring gives you a jump on some veggies by starting them indoors. Nightshades and

Brassicas are often started indoors. Some people skip this step and purchase seedlings.

- Covering seedlings on cold frosty nights in spring can protect seedlings, and allow you to plant them in the ground a little earlier. *Row cover* is a light fabric that is made specifically for this, but newspaper, boxes, bottles and jars also serve well. This can also help in late fall, but is more common in spring.

- Placing dark colored mulches can help warm the soil so you can plant earlier.

- Microclimate – Microclimates are small areas of the garden that are sunnier or shadier, wetter or drier than others. You can create your own microclimates to extend the growing season. Placing large rocks or stones in the garden can also warm the soil more quickly. In a strawberry patch this is near a concrete walkway, the strawberries near the walkway will bloom and fruit earlier. If a portion of that same strawberry patch is partly shaded, it will bloom and fruit later.

- Plant cool seasons crops several times during the spring and again during the fall. You can plant around the earliest recommended date and the latest recommended date, then plant once or twice in between. This will help ensure a steady supply of veggies.

- Shading cool season plants from the warm sun can extent their groing season.

- A cold frame is a box built on the ground. The bottom is just bare soil and the top is a glass panel, sometimes an old window or and old door. The ground will heat more quickly in spring and allow seedlings to be started a few weeks early. The glass top is usually hinged so it can be propped open for ventilation and for temperature regulation.

- A low tunnel is a row of hoops, one to two feet above the ground and a few feet wide, covered with row-cover or with plastic. It can add a few weeks on each end of the growing season, and is especially helpful for cool season crops. You open it to garden and to harvest from the outside.

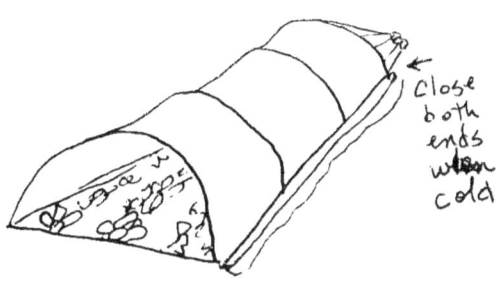

- A high tunnel is similar to a low tunnel, but is large enough to walk in. It is basically a low-cost greenhouse, usually not as permanent.

- Greenhouses are more permanent structures with lots of windows, or possibly a frame covered with plastic. It is convenient to build them against the house – to help keep them a little warmer. You can help keep it from freezing by adding thermal mass, in the form of barrels or jugs of water in dark containers that will absorb the heat during the day and radiate it at night. Some find that a few candles on the coldest nights are enough to keep it from freezing. Sealing the cracks can be important to maintaining warmth – but a tightly sealed greenhouse can get very warm in the bright sun, enough to kill your plants. If you can tackle the temperature regulation, with venting, thermal mass and some supplemental heat, you may be able to extend your season by four to six weeks in spring and again in fall.

Watering

Water is an important part of any garden. Many parts of the world are also facing shortages of reliable, clean water. For these reasons, your use of water in the garden should be done with care. Choose plants that match your landscape's water resources, drought tolerance, or moisture tolerance. If you match the plant to the site, you will need less supplemental water. In addition, mulching can help retain soil moisture. The same is true of building organic matter in the soil. (See Rainwater Harvesting.

Even with the best planning and practices, many sites will need some supplemental watering. Here are a few tips on making the most of your watering:

- Keep track of rainfall – it's a good idea to have a rain gauge, or even simply a jar placed out in the open on a porch or in the garden. This will give you a good idea of how much rain you have gotten, and how much you may be needing. Keep an eye on the weather as well.
- Another way to check for water needs is to look at the plants. If they are wilting, they may need water. Also, you can stick a finger into the soil – if it feels dry below the mulch and into the soil, then water may be in order.
- For fruits and vegetables, consistent moisture is key.

If the days are hot and dry, you may need to water deeply twice a week. Its better to water deeply every few days; shallow watering tends to encourage weeds.

- The best time to water is in the morning. That way the plants have a chance to dry more quickly and there is less chance for fungus to set in. Also, the water can soak in before the heat of the day evaporates it from the surface.

- Generally water early spring plants with a watering can, only when needed. There is usually less need in the early spring.

- When the strawberries blossom, start giving them supplemental water a few times a week with a drip line. Skip the watering if you have had a good rain. Extra water helps develop good plump berries.

- Extend the drip line to the warm season veggies: tomatos, cucumbers, melons, beans and sweet potatoes. Consistent water throughout the summer will keep these plants healthy and producing. I put my drip on a timer – every three days, they get a good soaking in the early morning. If there has been a good rain, I skip the watering session.

- Water fruiting trees and berries when it has been particularly hot and dry.

Rain Water Harvesting

Rain barrels have become a popular way of collecting water from the roof and using it to water our gardens. If you want to give this a try, be sure to calculate how much water your roof collects and ensure that you have a large enough rain barrel to collect all that will come down. One rain event can overwhelm standard 55 gallon barrels. That 55 gallons can then be used to water the garden. But again, it won't go far if you have a garden of any size.

After some experimentation, you may decide to give away the rain barrels and try storing rainwater directly in the soil. Work to build organic matter in the soil so that it was better able to absorb the water that came. That, along with deep rooted native plants, creates a storage cell for rain water that can handle a lot more than most rain barrels. It is also then available to the plants as they need it.

Another option for storing rain is a rain garden. This is a garden made of plants, particularly native plants, that are able to handle a wide variation in the amount of water they receive. They help to keep the water on site, rather than running off into storm drains. The rain garden acts as a sponge to soak up excess water and filter it through layers of soil into the ground water.

Choose a low lying site on your property, a place that usually forms standing water after a storm would be a good place to start. It should be away from buildings and underground utilities. Dig down a foot or so, so that you enhance water collection at your site.

Select native plants that tolerate standing water. Choose a variety of plants that bloom at different times, and that have different sizes and textures. Plant them in your rain garden and keep them mulched and weed free, especially in the first year.

The rain garden will bring beauty and biodiversity to your site. It will help to retain water on your property, which will enhance other plants as well. It will also help to enhance the environment and protect streams and rivers from storm water run-off.

Garden Markers

Often gardeners plant something, and promise themselves they will remember what is there. And then a few short weeks later the plant is growing vigorously, and they can't remember what it is to save their lives. Time and again this has happened to every gardener, and still they forget to mark a few things.

There are lots of commercial plant markers, sometimes they come with a plant if you buy it from a store. However, with plants grown from seed, or obtained from another gardener it is a challenge to find a way to mark them.

You can buy plant markers. You can make them from empty milk jugs or other containers. But the method that works best for me is to use a stick that's 1/2"-3/4" in diameter and about a foot long. Get a small bottle of green fabric paint with a writing tip on it. Write the name of the plant on the stick and stick it in the ground. It has a rustic look, and fabric paint is the one thing that doesn't fade or wash away. With perennials, longevity is important in plant markers.

Markers also help visitors and guests to identify plants, and even to help out with garden tasks. While the garden is very much home to the gardener, it can be a little like a jungle to the uninitiated. The markers seem to bring a little order to the chaos. And occassionally, they remind the gardener of something they forgot was planted.

Natural Building and Borders

Every gardener needs stakes, borders and pathways. Imagine growing those things in your own garden. Elderberry, mulberry, hazelnut, and willow are ready sources of stakes, and sticks that can be used for borders and chipped for pathways. Lots of plants require pruning, from berry bushes to fruit trees. Save these branches for your garden needs. You'll not only have a ready local supply, you will also solve the problem of carting away branches from pruning, and you'll have a reason to cut back an overly vigorous plant. One caution though. Make sure not to use any diseased plant material. This should be brough off site, preferably to somewhere that does hot-composting. Making stakes from your own garden plants adds a natural rustic look to your garden.

Natural pathways can be made from woodchips, and renewed every year or so with a fresh layer of woodchips. Plan the layout of the pathway. Lay a garden-hose along the path and adjust it to look right. Garden-hoses are great for this task because they can be used to give natural curves and gentle turns to a garden path. Once the path is planned, lay a good layer of newspaper or cardboard along the path. This can be placed right on top of turf and weeds. Put a few inch layer of woodchips on top of the paper or cardboard, along the whole length of the path. Over time, all the layers will break down into good rich soil. The path can be renewed with a fresh layer of woodchips every few years.

Wattle fencing is a woven fence that can be used for a variety of applications. It can be used as a garden fence. It can be used to form the frame of a raised bed or it can be used to establish terraces on a slope. Wattle fencing is a practice that is was used many centuries ago, they would gather sticks and branches to weave fences and borders.

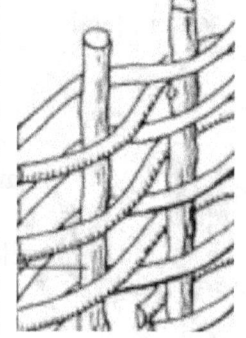

First gather stakes that will form the upright *staves* of the fence. The staves should be about one inch thick, hard-wood and tall enough do drive well into the ground and still reach as high as your fence will go. Then gather the horizontal *weavers* that are longer, thinner and more supple.

Plan the dimension of your fence and drive the staves into the ground, a few feet apart along the path of the fence. Next, place the first weaver in and out along the path of the fence. The next weaver will start on the opposite side of the sales and weave in and out. Continue to add weavers of varying sizes and lengths. Occasionally wrapping a weaver around the end and back into the fence to tighten the end. The whole project will look like a wicker basket that is woven right into the ground.

Borders can give a garden a more formal, finished look. This is particularly important when planting native plants or fruit trees in the front yard. Part of being a good neighbor is keeping up the appearance of the neighborhood. Native plants and fruit trees can appear unruly and may be an eye-sore for some. A crisp border will help to demonstrate the planning and effort that went into your beds. They also make it easier to maintain turf adjacent to the garden bed, and keep your mulch inside the bed.

Natural borders do all this, and also use natural materials that will add habitat to the garden. Stones are a common choice for a natural border. Another option is using logs and branches. Logs and branches can be gathered and added as the garden grows and evolves. Logs add to the natural-look of the space. If you gather the logs locally, they will be of various shapes, with the occasional side stump which can add character to the landscape. Gather logs of the same thickness, two to three inches is ideal. They will gradually decompose, providing added nutrients to the soil and habitat for critters who are visiting your site. You can add new logs as the old ones break down. No need to remove them, just add the new one as the old breaks down.

Hydroponics

Hydroponics is a way of growing plants in water with nutrients, without soil. Roots draw water and nutrients directly out of the solution. A net cup and a rocky medium (clay balls, perlite, rockwool) hold the roots in place. The nutrient solution flows over the roots and back into a storage tank, making this system extremely water efficient. There is no nutrient or water runoff.

Since the roots also need oxygen, the nutrient solution cannot be allowed to become stagnant. There are several ways to make sure roots remain aerated: aeroponics (where nutrient solution is misted onto roots), deep water culture (where an air stone bubbles the nutrient solution), ebb and flow (where the nutrient solution floods the roots and drains away periodically), nutrient film (where the roots grow into a thin stream of nutrient solution flowing down a sloping pipe), wick (where a wick brings nutrient solution from the tank to the roots), or the drip system (where nutrients are periodically dripped onto the roots - sometimes called Dutch Buckets).

Hydroponic gardens can be placed outdoors under the sun or indoors next to a window or under energy efficient LED grow lights. Indoor hydroponics give a year-round harvest that is protected from birds, bugs (you'll never even be tempted to spray a bug), and bad weather. This means your food can take mere minutes from harvest to table any time of the year. Outdoor hydroponic gardens thrive even on the hottest sunny days, because the plants won't dry out like a container or soil garden would. Another benefit of hydroponics is leaving the soil undisturbed. A raised hydroponic garden can provide partial shade to perennial crops grown below.

Many common garden plants thrive in hydroponics. If you buy local produce in winter, you probably already eat hydroponic tomatoes, lettuce, and strawberries. Cucumbers, peppers, and herbs also grow prolifically when provided the perfect supply of water and nutrients, which hydroponics allow.

Commercially available nutrient mix is the easiest way to begin growing hydroponic produce. Simply mix according to package instructions and replace the nutrient solution in the reservoir after it is depleted or every several weeks. Advanced gardeners who wish to move away from synthetic fertilizers can use carefully prepared compost tea as the nutrient source, or raise fish, which provide nutrients in their excrement. Growing plants in conjunction with a fish tank is known as aquaponics.

Plant Propagation

Plant propagation is the name we use for a variety of methods of increasing the number of plants we have. In plant propagation, we make use of a plant's natural capacity to reproduce itself, and manage it for our own needs. Some plants reproduce very readily and we have to work to contain them in the garden. Other plants need a little encouragement to reproduce where and when we need them.

Plants reproduce through flowers and fruits (called sexual reproduction) and through spreading of the plant itself (called asexual or vegetative reproduction). We'll talk first about sexual reproduction, involving collecting, storing and planting seeds. Then we'll cover asexual reproduction, through softwood and hardwood cuttings, root cuttings and divisions.

Seed Saving

Plant flowers contain male and female parts. Pollen is transferred from the anther to the stigma of a flower and fertilization occurs. Some flowers are 'complete-flowers' containing both anther and stigma, and thus the flower can self-pollinate. Seeds from self-pollinated plants will have the same genetic make-up as the parent plants. Sometimes pollination is done with the aid of wind or insects. Plants with large showy flowers are usually pollinated by insects or animals. Wind-pollinated plants often have small, even insignificant flowers, for example, grasses and some trees. Some flowers have separate male flowers and female flowers, and sometimes, the male flowers and female flowers grow on separate plants. It is helpful to learn the reproductive characteristics of the plant before collecting the seed.

Once pollinated the seeds begin to form, sometimes within a fruit, as is the case with berries and pawpaws, tomatos and melons. Other times, the seeds form in a pod as with legumes, or in the open as with asters, milkweeds and lettuces.

Observe plants growing in the home garden, watch the flowers and fruits develop. Pull a flower or fruit, now and then, to assess the maturity of the seed. The Seed Savers Exchange (www.seedsavers.org) has information about saving many species of seeds. Here are some general tips:

- Choose – Plan which seeds you will be collecting. Native plants are usually open-pollinated, so they are a good place to start. Heirloom vegetables are also open-pollinated. Hybrids show wide variation in the plants that grow from their seeds (F2 generation), so aren't good choices for seed saving. Seed Savers has lots of information about easier and harder seeds to save.

- Time – Keep an eye on a plant as the flowers die, and/or the fruit ripens. In some cases, it can be helpful to place a mesh bag over ripening seeds so that they don't disappear before you have a chance to collect them. You may want to mark a plant when it is in flower, so you will recognize it when the seeds ripen. Get permission to collect in public spaces or in others' gardens.

- Gather – Make sure you gather ripe seeds. Collect them in used envelopes, or folded sheets of paper.
- Label – Label each type of seed with the plant name, place and date of collection. Don't rely on memory.
- Dry – Allow seeds to dry thoroughly, either in a paper envelope, paper bag, or on an open surface. They should be in a dry place with good air circulation, without insects, birds or mammals.
- Clean – It is not necessary to clean the seeds from the rest of the plant matter (the chaff), but this step can help to make seeds easier to store and keep seeds fresher. Cleaning seeds is a good indoor activity for fall and winter, when gardening tasks are at a minimum.
- Store – Seeds are living, though they are dormant. For this reason, it is important to give them the best conditions to stay alive. Once they are fully dry, store them in air-tight containers, put all your envelopes in a canning jar. Or separate them by species into old pill bottles or spice bottles. Keep the seeds cool, but not frozen.
- Treat – Some seeds require 30-90 days of cold treatment (stratification), placing them in the refrigerator will meet this need. Some seeds have a tough seed coat that has to be broken before the seed can germinate. This can be done just before planting by nicking them with a knife or sandpaper, or by soaking them in water.

- Share – Collecting enough seeds for the home garden is enjoyable and cost saving. Expand your varieties by bringing your seeds to a seed-swap, sharing with others and collecting the seeds they have to offer.

SEED STARTING

Most native seeds can be planted outdoors in late winter. They will benefit from the variations in temperature and moisture, and will germinate when conditions are right. Native plants are adapted to spread their seeds without human intervention. Observe when the seeds ripen and how they are dispersed. Follow that pattern for best results.

If you are starting seeds indoors, make sure to give them sufficient resources to thrive. See the section on Air, Earth, Fire and Water. Most seeds germinate best at around 70-degrees. Keep seeds moist until they develop a root system to absorb moisture from the soil. Once sprouted, give the seedlings plenty of light and some moving air. Young seedlings are susceptible to fungal infections called dampening-off. The best prevention is a healthy growing space.

Many growers insist on using sterile planting mix. Others get great results by using soil straight from the garden, especially if they have been working for years on improving the soil microbiology and organic matter.

Start plants 4-6 weeks before the last frost. Slow growing natives can be planted even earlier.

Some find it helpful to plant 10-20 seeds in a 4-inch pot. Then in 4-6 weeks, when the seedlings have several true leaves, they can be gently separated into individual pots. It is easier to pamper one 4-inch pot than 20 individual pots.

When transplanting to larger pots or transplanting into the soil, pamper the plants for a few weeks. Make sure they have protection from too much sun, too little water, and from critters who would prey on them before they get established.

CUTTINGS

Many plants can be propagated by taking cuttings from an existing plant. Cuttings should be pencil-thickness. Cut at a node, i.e. where a leaf is attached. Cuttings should have 4-6 nodes. Keep cuttings moist and place them in growing medium as quickly as possible after cutting. Take extra care of cuttings until they are well rooted. If you put them in pots, put them in a protected spot to root and grow out, before planting them out in the ground.

- Softwood cuttings are taken in late spring or early summer from plants that are actively growing. Leave only a few leaves at the top of the cutting. Place it a pot in moist soil. Cover the pot, or mist the cutting regularly until it roots. It should take a few weeks. You can dip cuttings in rooting hormone, or water them with tea made from willow stems. Sprinkle the cuttings with cinnamon to prevent fungus.

- Hardwood cuttings are taken from one-year-old wood as the plants go dormant in the fall. Remove all leaves, and place them in soil to over-winter. In the spring, it may have formed roots and may leaf out. Keep it protected through the first few months in spring until it is well established.

- Root cuttings are taken from plants in the spring, especially from plants that send rhizomes (underground

- runners). Plant these in the ground and keep them well watered. They should send up shoots in 6-8 weeks.
- Layering is a method of rooting a stem without cutting it from the parent plant. Bend a stem down to the soil and secure it there with a rock or a landscaping staple. Over time, the stem will form roots and can be cut from the parent plant. An alternation of this is to put a handful of soil around the stem of a plant, secured by plastic wrap or a plastic bottle and kept in place with string or tape. Keep the soil moist and cut from the parent plant when you begin to see roots emerging into the soil.

Division

Some plants grow in clumps that expand from year to year. These plants can be propagated by division. As a general rule, plants that flower in spring and early summer should be divided in late summer or fall. Those flowering in summer and fall should be divided in early spring before new growth begins.

Dig the plant up, and remove enough soil to get a good look at the root structure. Divide the plant into clumps, keeping several shoots and roots on each clump. Sometimes there is an older dead portion in the middle of the plant – that can be discarded.

Shrubs can also be divided when they are dormant. Cut the shrub back to one or two feet tall and dig it out of the ground. Dividing the shrub may require a saw or hachet, keeping several shoots and roots on each clump. Put each clump in a separate pot or in a separate spot in the landscape.

Animal Life

Plants aren't the only part of the landscape, animals are a vital part as well, helping to recycle nutrients, and providing benefits to the landscape in the form of ecosystem services. Let's look at a few in particular.

Pollinators

Pollinators are essential to a healthy ecosystem and they are essential to many of our food crops. There was a time when we could take our pollinators for granted, but today, many of our pollinators are in decline due to pesticide use, loss of habitat and climate change. It is important for sustainability that we do what we can to support pollinators. We can do that through several practices that are also important for growing flowers and for producing fruit and vegetables in our home gardens.

- Nectar - Provide pollinators with food sources in the form of flowers. Ideally, you'll have a few species of flowers blooming throughout the growing season beginning in early spring. Don't forget flowering trees that bloom in spring and provide a good start for the pollinators.

- Cover – Provide pollinators with places to hide in bushes and other vegetation. Some pollinators also overwinter in the native plants and leaf litter that remain in the garden. Leave some of your garden clean-up tasks to late winter or early spring to give the pollinators a place to stay. See the section on Garden Clean-up.

- Nesting – Provide pollinators with places to build their nests. Most of our native bees nest in bare ground, so make sure there are places between your plants that aren't mulched. Some also build cavity-nests in old twigs, branches and stems. You can leave some brush for them, made of flower stalks, branches pruned from your berry bushes and a few larger logs. You can build or buy bee-house, but make sure you are careful to maintain it so that they don't spread diseases.

- Pesticide-free – Ensure that there are no pesticides used in your garden, and if possible, provide a buffer if your neighbors use pesticides. See the section on managing pests to learn about gardening without pesticides.

- Water – Finally, provide your visiting pollinators with a source of water such as a bird-bath or a small garden pond. Make sure the water is clean and doesn't breed mosquitos.

Birds

Growing bird-friendly plants will attract and protect the birds you love while making your space beautiful, easy to care for, and better for the environment. The birds interact with the ecosystem to improve its biodiversity.

Chickens, ducks and quail can also be grown in a home garden and they add to your system as well.

Pond

A small pond is a great way to add biodiversity to your garden and offer a ready supply of water to all of the critters who visit. Ponds can be large, elaborate and costly. Or they can be quite simple and easy to maintain. Here are simple elements in a small pond ecosystem:

- Choose your site – preferably one that is out of the way of traffic, but close to the patio or to seating. You'll want to observe your new pond and all the wildlife it brings.

- Dig a hole and choose a liner. One easy way to do this is to choose a small rubber barn-yard trough – 20 gallons capacity will do the trick for a small garden pond. This won't be deep enough for fish to winter over in my climate – so it will have to be dumped, re-planted and re-stocked each year.

- Select a few pond plants that are native to your area. They are likely to survive from year to year in your climate. Possible choices include Soft Reed (Juncus effusus), Arrowhead (Sagittaria latifolia) and Bog Bean (*Menyanthes trifoliata)*. Stepping away from the native plant theme, Azolla makes a great cover for your pond. It fixes nitrogen, gives the fish something to munch on, and spreads well. If a garden plant needs a little boost, a scoop out a handful of Azolla and put it at the base of the plant covered by a rock, or some leaf litter.

- Fish are a great addition to the pond, a few minnows are a good choice. They are native to our region and are fairly hardy. They will feed on mosquito larvae and on the pond plants – no extra feeding needed. Before you add fish, make sure to air off the water for a few days – our water has chlorine. It's also a good idea to get a small jar of water from a local fresh water pond or lake. Add that to your pond to jump start the microbiology that will help to process fish waste.

- If you don't have fish, you'll need some sort of pump to keep the water moving and avoid breeding mosquitos.

- The pond should have shading to keep the water from getting too hot on bright sunny days. Also, if the pond gets too much sun, algae can take over the whole space.

- You will probably need to top up the water periodically over the summer. Make sure to air off the water you will be adding – and try not to add more than 10% of the pond volume at a time. Rain water is a great way to top up a pond. Let it rain in directly. Or collect the rain water in a bucket and add it to the pond.

- Sit back and enjoy the critters as they enjoy your new space.

URINE

In considering the benefits of animal life in the landscape, don't forget the human-animal life. We too give off waste products that can be useful in the landscape. Think of all the critters the both poop and pee in the woods and fields, and the plants thrive on recycling the nutrients. Urine in particular is rich in nutrients needed for healthy plant growth. Poop is difficult to treat, but there are resources that can help you to do it safely and relatively sanely. The same is true of poop of humans, dogs and cats. But urine is sterile in a healthy person, and completely safe to use in the landscape, with a few precautions.

- Urine should be collected from a healthy adult. Someone with an infection or someone who is taking pharmaceuticals would not be a good candidate.

- Generally urine shouldn't be put directly on plants, and especially on food plants. However, it can be put in the mulch before a veggie garden goes in, or in the mulch layer around a tree. It can also go on the compost pile, on woodchips or leaf-piles.

- For a small urban lot, one person's urine is probably all the system can handle. This much can be applied though out the system, along with regular watering.
- Some dilute the urine, some apply directly – this is more about ick-factor than about safety or plant nutrition.
- Let your nose be your guide. You shouldn't smell anything other than rich soil. If you still smell the urine a day or two after application, you're probably over-doing it.
- Collection can be done inside – think chamber pot. Then brought out and applied, straight or diluted. Make it a point to apply in different places each time, moving around the landscape top-dressing home-collected fertilizer.
- An added benefit is that you pull some of this waste out of the sewage system, saving clean water in the process.

HOME

This chapter will discuss various aspects of sustainability in the home under three main headings: Food, Energy and Household issues.

FOOD

We'll start with food, this follows up the first chapter on gardening, especially the concern with producing food right in your backyard. This section will discuss what happens to the food once it reaches the door of your home.

ORGANIC

Organic foods are foods that are raised without the use of synthetic pesticides and synthetic fertilizers. Much of conventional agriculture includes the use of chemicals to kill bad bugs and chemicals to enhance soil fertility. However, these chemicals and fertilizers also degade the entire ecosystem, build up in soils and remain on foods that come to our supermarkets. Organic farmers must comply with certain standards to be certified organic. Many home gardeners go far beyond these standards in their use of natural means to grow food. Ecovillagers aim at gardening practices that are sustainable, and even practices that regenerate the ecosystem, leaving it better than when they arrived. There are several reasons advanced for growing and eating organic:

- Nutrition – Dietitions report that foods grown organically are more nutrient dense than conventionally grown foods. Consumers report they are more tasty.

- Chemicals – Chemicals used in conventional agriculture remain on foods, and in some case inside foods that are brought to market. It is difficult to remove all these chemicals in food preparation.

- Biodiversity – Organic gardening and farming practices support greater diversity of insect life both in the soil and above the ground.

- Soil – Organic gardening and farming practices support better soil. Chemical fertilizers tend to decrease soil structure, which leads to the erosion of top soils.

- Pollinators – Avoiding pesticides in organic farming and gardening helps to support the health and abundance of native pollinators and the overall health of the ecosystem.

If you grow foods in your back yard, try to reduce or eliminate your use of synthetic pesticides and fertilizers. If you buy your food, consider spending a little more for organically grown foods, to support the farmers who are choosing more sustainable practices.

Plant-Rich Diet

Some choose to forgo meat entirely, and eat vegetarian, others choose to avoid all animal products, which is vegan. It may be more realistic to see this as a spectrum. Those who advocate a plant-rich diet encourage people to increase the foods that come from plants, thereby reducing their consumption of meat and dairy. This can go a long way to reducing the emissions because the meat, dairy and poultry industries have high carbon emissions.

This is challenging because food is personal and cultural. At the same time, it can be a simple as having one all-plant meal a week, then adding a second and a third. As we learn new recipies and establish new traditions, we can be helping to save the planet as well.

Glass Jars

Store food in long lasting glass jars. They are more hygienic, and environmentally friendly. A friend chooses to buy jellies and sauses in glass jars and saves those jars for food storage. Over time she has a growing collection and has many canning jars alongside the branded jelly jars. These jars can be re-used indefinitely. Occasionally she has to buy new tops for the canning jars. And as she get's more of those standard canning jars, she gives the others away – often with some tastey home-grown offering.

Sun Tea and Solar Cooking

Sun tea is made by putting fresh herbs or tea-bags into a jar and placing that in the sun for a few hours. The sun heats the water and steeps the tea. Add flavorings or sweeteners, and serve over ice.

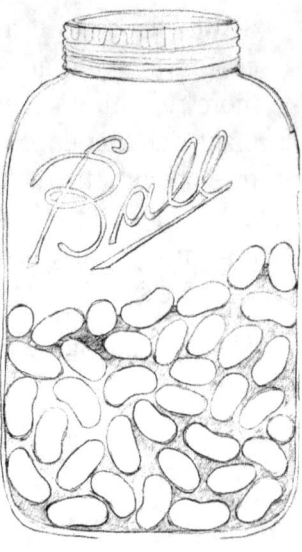

Some use this same technique to do summer cooking, for example, soaking beans in the sun. Put dried beans in a jar, covered by plenty of water. Beans will expand, but won't cook entirely. However, this method will cut down on cooking time, and on energy used. Smaller beans and grains can be completely cooked using this method. Try quinoa, lentils, rice, noodles or split peas.

Take this same method up a notch by painting jars black, and/or placing them in a three sided box covered with mirrors or another reflective surface. Actual cooking time will vary with the amount of sunlight and wind, and the outdoor temperature.

Camp Coffee and Cold Brew

Many homes feature single serving coffee-makers – with lots of plastic waste with each cup. Older drip coffee makers are better, but require continual purchase of filters. Our grandparents had a coffee pot that heated on the stove - no waste. A french press is a small glass pitcher with a wire mesh plunger. Put in the coffee and boiling water. Once it has steeped, push down the plunger and pour out a fine cup of coffee.

And then there is what my friend calls "camp coffee." Throw some grounds in a cup, add boiling water and give it a stir. It's the first thing she does in preparing her breakfast. By the time everything else is ready, the coffee has steeped and the grounds have mainly fallen to the bottom of the cup. She drinks it – but not 'to the dregs'. Once she gets to the bottom, she pours the last of it on a plant that could use a fertilizer boost (coffee grounds are rich in nitrogen, a key component of fertilizer). She has her boost for the day, and offers a boost to the garden as well.

Cold brew coffee is another variation that saves the energy of heating the water, and avoids putting heat in the house, especially during the warm summer months. It also has a distinctive flavor, sweeter and gentler than regular hot-brew coffee. Choose a large glass jar, enough for your day's supply of coffee. Put the standard amount of coffee grounds in the bottom of the jar, a 1:8 ratio of grounds to water works well. Fill the jar with water, place the cap on tightly and gently shake till the coffee and water are mixed. Sit the jar on the counter or in the frig to steep for 12-18 hours. Use a fine mesh strainer to strain the coffee into another container to remove the grounds and then it strain back into the original jar. It will keep for weeks in the fridge, or days on the counter. If you prefer hot coffee, you can always heat it before drinking.

Preservation: Solar Drying

Solar drying can be an easy, low-cost method of food preservation. Once you have dried food, its sides are greatly reduced. It can then be stored in the freezer to extend its life. Most of the time the abundant harvest comes at a time when there is a lot of sun to help with the drying, and you don't have to put extra heat in your home, as you would with other forms of preservation. Drying also preserves more of the nutrient content in the food.

You can often create a solar drying system out of things you have around the house and garden. You can place veggie tomato slices on a tray covered by a cloth towel to keep insects away. Put the tray in your car, parked in the sun on a hot day. You could place the same tray in a greenhouse if you have one standing empty during the summer. Those who do a lot of drying may set up something more permanent. Commercial solar dryers are available, there are also lots of diy plans in books and on the internet. In between the parked car and the commercial models, there are some semi-permanent setups that are convenient and can dry a lot of food. Keep in mind the following when setting up your drying station:

- Exclude insects, leaves, etc. Mosquito netting or window-screen can be used for this.
- Concentrate solar energy, either by making the container black to absorb heat, and/or having a clear covering to trap solar energy, like a greenhouse does.
- Allow for air flow to bring the moist air away from the food, in a sealed container, the moisture will condense on the frame.
- Make trays for the food. They should be permeable to allow for air-flow, non-stick and easy to clean.
- Allow enough time to dry – depending on the food, the efficiency of your system and the temperatures, it may take several days. Allowing the partially dried food to cool overnight can cause the food to mold. Some prefer

to us an oven or electric dehydrator to finish off drying and to preserve quality and food safety.

- Slice foods to allow for quicker drying.

Store dried foods in a sealed container. If they are completely dry, they stay fresh longer. If the food has a leathery consistency, it's best to keep it in a sealed container in the freezer.

Preservation: Fermentation

For much of human history we have used fermentation to preserve food and prolong the shelf life of foods we harvest or forage. It is a magical process that transforms food. Two primary means of fermentation are salt-based fermentation of vegetables and sugar based fermentation of fruits and berries.

Salt-based vegetable fermentation

Lacto-fermentation is a fermentation process driven by lactic acid bacteria, which convert sugar into lactic acid. This is done by submerging vegetables in salty water, called a brine. The salty brine creates the conditions for good microbes to grow.

Many vegetables can be fermented in a salty brine, with a variety of herbs. We also have to be careful in introducing our home fermented foods to our diet.

If your water has chlorine in it you can boil the chlorine off, or just let the water stand on your counter for a day. The chlorine will inhibit the growth of bacteria, so you want to eliminate it before starting a fermentation.

A simple recipe for fermentation would be to cut pack chopped vegetables into a quart jar and add a tablespoon of salt. A few herbs can also enhance the flavor. Put a small weight on top of the veggies to hold them under water. Let the veggies ferment at room temperature for a week or so. Then lightly cover the jar to keep the critters out and still allow air flow. The flavor will begin to develop in a few days and will continue to mature over the next few weeks. Taste the mixture every few days and when you like the flavor you can eat it, or put it in the refrigerator to slow the fermentation.

There are lots of great resources on fermentation. I find the best resources at #wildfermentation or anything by Sandor Katz.

Simple Country Wine

Fruit fermentation is simple enough: fruit, a sweetener, water, and a starter culture if you have one. For a lightly fermented berry wine, mix together in a large canning jar:

- 1 part fruit
- 4 parts un-chlorinated water
- 1 part sugar
- 1/4 part lemon juice to lower the pH
- 1/4 part starter culture (optional), e.g. a tablespoon from your last batch.

Float a plastic lid on the top to cover most of the liquid and cover lightly with a cloth to keep out bugs and allow air exchange. Stir several times a day.

You will see bubbles forming in the mixture, add sugar if it stops bubbling. Taste the mixture every few days and when you like the flavor you can drink it or bottle it. I decant the liquid into flip top wine bottles, add a little extra sugar and seal them. Refrigerate after a few hours at room temperature. This sealed fermentation will add fizzyness to the wine, so use caution.

HOUSEHOLD

There are lots of household tasks that can be done in a more environmentally friendly way. We have come to use so many disposable convenience items around the house. We have come to use so many toxic chemicals. Let's tthink about keeping house the way our grandparents did. They had many tools and tips to keep the home clean while protecting the environment.

REDUCE, REUSE, RECYCLE

Really, it is as simple as all that. Reduce home size – and you reduce heating, cooling, cleaning, furniture, etc. Reduce clothing, and you reduce storage, cleaning, waste, etc. Reduce food inventory – how many of us could survive for months just with the food that is already in our homes, yet we go out regularly to shop for more – and have to store, and eventually discard what we have.

One friend has committed to discarding, recycling or giving away two items of clothing for every one she acquires. Another friend only keeps clothes that she has worn in the past year. If she is changing out seasonal clothes and realizes that there is something she hasn't worn, then out it goes, and it's not replaced.

Re-use is another goal. Re-usable grocery bags have become a bit controversial. Some people say too many of them are used infrequently and then discarded. One friend has a cloth bag made by her mother who has passed away. She has used the bag for more than five years and it is still in great shape – being well made in the first place. Another friend has sworn off single use products: paper plates and cups, and even more plastic cups and forks and spoons.

And finally, when something can no longer be used, make sure you know how and where to recycle it. Most cities have curbside recycling, it just takes an extra step to separate and clean items to keep them out of the landfill.

Many of the suggestions in this book will also reduce the disposable items you use. See Cleaners, Cleaning Supplies, Gardening, etc. Gardeners often remark how little waste they have since they are not bringing produce home from the store in all the packaging that comes with it.

Cleaners

Many commercially available household cleaners contain chemicals that are harmful to the environment. They come in bottles and containers of all sorts, and we seem to need more and more specialized cleaning products, expanding the number of bottles in our pantry.

Most household cleaning can be done with baking soda, vinegar and borax.

Vinegar works great on windows – when I was I kid we cleaned windows with vinegar and old newspapers to dry the windows and get the streaks out. Vinegar also makes a great floor cleaner, cutting grease and grime and leaving the floor just as squeaky clean and the windows. I put some vinegar in the pail of water and mop away. Vinegar is also good for counter tops, sinks and toilet bowls. Vinegar and water is pretty much an all purpose cleaner around the house.

Baking soda is good as an oven cleaner. Sometimes I switch it up and use baking soda for something I usually clean with vinegar, e.g. the floors. I figure it cleans a different way, so I'm hitting different kinds of grime.

Baking soda also works for shampoo. I put a tablespoon or so in a cup of warm water. Pouring it over my head, I massage it into my scalp. You'll find it's a different kind of clean from shampoos that remove all the natural oils. If you like you can follow up with a little regular shampoo, or go with baking soda only. Vinegar in water makes a good rinse.

Baking soda also works for toothpaste. It cleans your teeth and leaves your mouth feeling fresh. I found that I am much less susceptible to dental cavities because it neutralizes the acid in my mouth.

Recently, when loading the dishwasher after a potluck, I found we are out of dishwasher soap. After I quick search online, I found the recommended substitute: one tablespoon of borax and one tablespoon of baking soda – with vinegar as the rinse-aid. It worked like a dream and I never bought dishwasher detergent again.

Borax is another great household cleaner. It's great for the bathroom, sprinkle it around and scrub with a cloth or brush. You can use the same cleaner for the kitchen. Again I switch it up with vinegar and baking soda on counters, floors and appliances. Borax also works for scrubbing pots and pans.

It's probably a good idea to start small, make one substitution that makes sense to you. One by one, you can switch your cleaning routine to these simple, cheap, envirnomentally friendly products. The next time a cleaning solution runs out, before running out to buy another bottle, consider it's "old-fashioned" replacement. The internet can be helpful in learning these substitutions.

DIY Laundry Detergent

You can make your own eco-friendly laundry detergent with the following recipe. There are lots of other recipes floating around, but this is the one that works best for me. Mix the following in a large mixing bowl, then store in quart canning jars and label accordingly. Use a tablespoon or two for each load of laundry.

- 2 parts Baking Soda
- 2 parts Washing Soda
- 1 part Epsom Salt
- 1 part Castile Soap grated (optional)
- a few drops of essential oil (optional)

Ants and other Household Pests

Household ants are a part of life. Naturally, keeping things clean, keeping food tightly sealed and cleaning up promptly after meals can keep ants at bay. But they seem to make their seasonal appearance, despite best efforts at cleanliness and tidiness.

So when the ants come in, I have a trusty remedy. If it's just one or two, I can over look that. But if they bring the family, then I go for sugar-borax-water solution.

I take a small glass jar – mine once held a jelly sample. I clean it out and take off the label. Then I mark "sugar & borax" on the jar with fabric paint – my go-to for permanent marker. Then I put about a teaspoon of sugar and a teaspoon of borax and close tightly. This is usually enough to last the season. I save bottle tops and soda bottle lids back from the recycle.

When the critters come along, I take out a bottle cap and put about ¼ to ½ teaspoon of the sugar-borax mix in it. I add a little water and sit the bottle cap with the borax-sugar-water solution in the path of the ants, but out of the way of the people. The ledge at the back of the kitchen counter works perfectly, and in another case, the corner of a window-sill.

It takes a 3-4 hours for the ants to find the solution, and they will tend to mass around it. In another 6-8 hours, they will be gone. I then toss out the whole bottle-cap with solution and ants. I might have to do this a few times before they are gone.

It could be more earth friendly if I used something compostable to to hold the solution. That will be my next step.

Now that you know how to handle ants, let's think about some of those other critters that try to invade our space. Certainly keeping our homes clean and relatively free of clutter will go a long way to reducing all sorts of unwanted visitors. But the occasional roach or spider will probably venture in despite our efforts.

I tend to trap spiders under glass jars, and remove them to the outdoors when I have a chance. I slip a piece of stiff paper or thin cardboard under the jar, and carry the critter outdoors for release.

I'm less forgiving with beetles and roaches. They can generally be stopped in their tracks with some soapy water. And that's less of a clean up issue than other bug-killers. Borax-sugar-water used for ants can also be used against these pests, but it's a little slower to act.

Flying/stinging pests are a challenge. But before you reach for some toxic spray, consider trapping the bee or wasp in a towel. I learned this one from my Dad. He would take several thicknesses of towel and approach the critter stealthily, quietly grab it in the towel, and walk outside with it buzzing away in the wad of towel. Dad just tossed the whole thing outside and let the critter fly away. No harm, no foul. He would retrieve the towel later.

Cleaning Supplies

Cleaning supplies have also become more plastic, more disposable and more toxic. Think back a few generations and maybe you can bring back good "old-fashioned" clean to your home. Remember these cleaning supplies? If not, maybe your parents or grandparents do:

- A straw broom with a wooden handle. It lasts for a good long time, and when it's useful life finally does come to an end – you can use the pole for a tomato stake and compost the straw part.

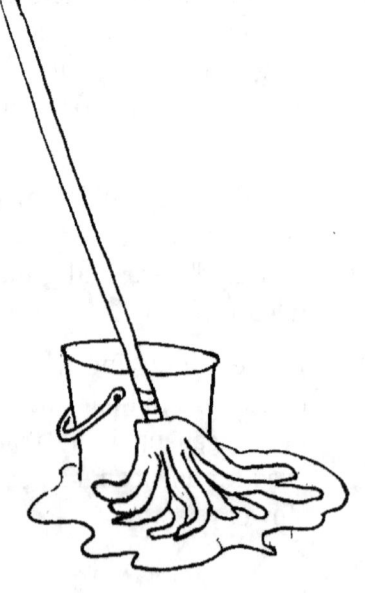

- Remember cloth rags made from old t-shirts. They can be washed and reused for a long time. And you'll save lots of paper-towels and avoid micro-fiber products that clog the environment with tiny plastic particles. A cotton mop is a great cleaning tool and can be washed again and again – and while your at it, a good sturdy bucket will help you get your cleaning tasks done quickly and efficiently. And you can throw that dirty water outside to water a fruit tree, or those native plants.

- See Cleaners – for a list of cleaning products that are cheaper and safer for the environment.
- A scrub brush can be a great tool for attacking tough dirt or built-up scum.
- Work clothes are a great help – that way you don't mind getting down and dirty to get your home tidied up and clean. They can go in the laundry with all those rags.
- And don't forget a good playlist to keep you going and a shower and a cool drink at the end of the task. Well done!

Basketry

Basket weaving started back in pre-history when people wanted to tote things from place. It enabled people to carry foraged berries and mushrooms, supplies or tools, with a basket made from plant materials that grew around them. We can replace some of our plastic bags and totes with home-made baskets made from natural materials.

Baskets are hand woven from plant materials, using a variety of techniques that are grouped into three broad categories: coiled, woven, and twined. The materials range from willow wands to pine needles and include grasses, wood splints, palm fronds, reeds, brambles and bark.

Woven

Woven baskets are often made of tree bark, which can be stripped of fallen or pruned trees in the spring and summer. The bark is cut to even widths for weaving.

- Choose twelve stakes that will go across the base of the basket and up two sides. In this example, let's use a 4" round base and 2" sides, so you'll need 4"+2"+2" plus a few extra inches to work with and weave in the ends.

- Lay down six strips of bark parallel to one another, making sure they are long enough for the base and two sides with extra to trim. For this example, we will make a four inch square base and two inch high sides. Use a weight, clip or tape to hold the first strips in place.

- Weave a second set of six bark strips in and out through the center of the first strips. You will end up with a four inch square of woven bark in the center with extra bark strips extending evenly on each side.

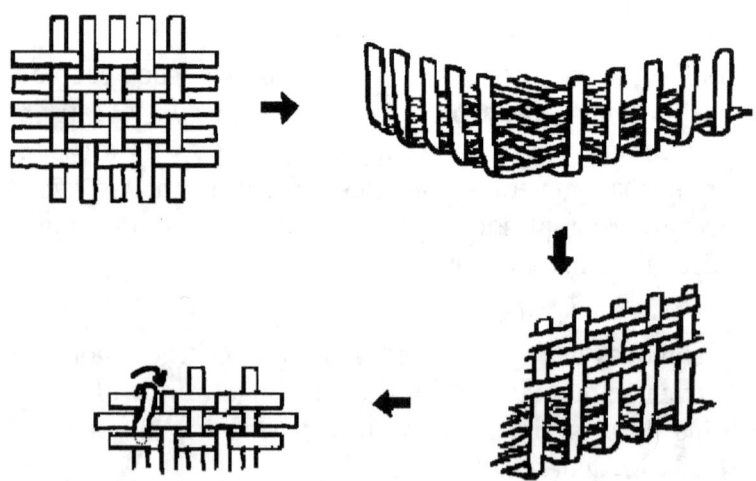

- Once the center is complete, it is helpful to weave a piece of twine in and out around the woven square to hold them in place before proceeding.
- Upset – Gently fold each of strips of bark upward. They are now the stakes for the side of the basket
- Weave a strip of bark, over and under around the basket, overlapping the ends a bit to hold them in place.
- Weave additional strips, alternating over and under, with the previous round. If the 1st strip goes under, the next strip should go over.
- When you reach as high as you want to go.
- Finishing – Working with every stake where the top row of weaving is on the inside of that stake, gently bend the stake at the edge of the top row, cut it to length so that it will be hidden by the third row of weaving (from the top), and tuck the stake under the third row on the inside.
- Cut the remaining stakes even with the top of the basket.

Twined

Twining is a simple and flexible technique for using a wide variety of softer materials. Harvest materials and fully dry them. Then wet them just enough to make them workable.

- Choose eight stakes that will go across the base of the basket and up two sides. In this example, let's use a 4" round base and 2" sides, so you'll need 4"+2"+2" plus a few extra inches to work with an weave in the ends.

- To make the base, lay four stakes north-south on your table, and lay the other four east-west over them.
- Choose smaller more flexible materials for your weavers.
- Weave over the east, under the north, over the west, under the south, and repeat, going around a few times until the stakes are secure

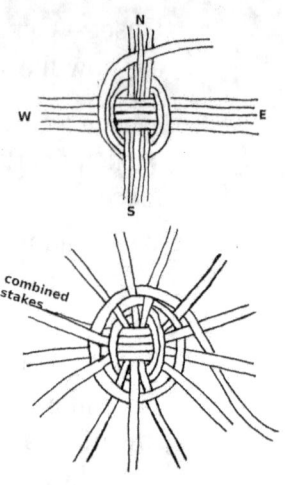

- Spread the stakes out evenly around the circle - i.e. at all points of the compass.
- Combine any two stakes that are side by side so that you end up with an odd number of stakes.
- Twine over one stake, under the next, over, under, continuing around until you have a 4" circle – according to your plan.
- Upset – Gently bend each of the stakes up, this is called the "upset". They won't stay up until you begin twining up the side, pulling tight to hold the stakes up.
- Sides – Weave up the side – it will get easier once the sides are established with a few rows of twining. When you've completed a round or two it's good to pull all the stakes nice and tight and neaten up the weaving up a bit.
- Joining – Whenever the weavers are getting a bit short or becoming too uneven in thickness, it's time to add a new one. Lay the new one alongside the old and twine

them together on a few stakes, then continue with the new. You can trim the tail of the old after a bit.

- Continue to twine up the sides, keeping it tight and neat as you can and shaping it the way you like it.

- Finishing – Starting with any stake, take it around the inside of the stake directly to the right and back to the outside of the basket, down and out of the way. Don't pull this first stake down too tight as you'll need to thread the last stake through at the end.

- Moving to the right around the basket repeat this action, folding each stake over the stake to it's right, tucking the last stake through the original loop then pull tight.

- Once you have finished, clip any loose ends. If the basket is wobbly, while it is a little moist, put it on a flat surface with a few weights in it to hold it stable.

Coiled

In a coiled basket you wrap your core material into a coil and 'sew' over the coils with another material (the stitching material) with a needle.

The core material can be natural, such as grass or pine needles, or it can be household materials such as rope, or recycled paper or fabric scraps.

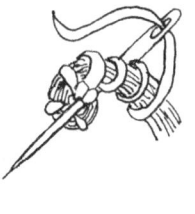

The stitching material can be natural plant fiber or it can be cotton string strong enough to bind your basket together.

Form a tight coil inner circle of your core material and place a few stitches through the center and around the outside of your inner circle.

You can continue to wrap your stitching material tightly around the coil material, stitching after a consistent number of wraps. Alternatively, you can just place a stitch every 1/2 inch, continuing to coil the material outwards until the coil is large enough to form the base of the basket. Start building the side of the basket by stacking a coil on top of the previous round and stitching it in place.

When you are finished, taper your coil, stitching it so that it blends into the previous layer.

Making Cordage from Foraged Fibers

Most of human history, people made cordage (rope or string) from the plants that they had around them. It can be a fun and rewarding skill to develop. You can also replace many home and garden string needs with cordage fashioned on the spot.

Harvesting Fiber

First, you'll need to identify plants with long, strong fibers. Common plants for cordage include cattails, nettles, milkweed, dogbane, and yucca.

Fibers require different harvesting and preparation techniques, google the specific fiber to learn more. For example, milkweed, dogbane and amsonia are best harvested in winter after a good freeze. Lightly scrape the stem with a thumbnail to flake off the outer skin. Crush the stem in half, then splay the stem open lengthwise. Break a few inches of the woody material and lift it from the outer fibrous material. Work your way up the stem, removing the pith and woody material and you will be left with a ribbon of fiber with some plant material still attached.

Yucca requires a different treatment. Cut the green leaves from the plant and roast them over a fire till the edges are singed. Then pound the leaf on a firm surface with a smoothe stone. You will see the fibers start to emerge. Through pounding and scraping, you can gradually refine the strong smooth yucca fibers.

Once you have some fibers, make sure they are completely dry before working them. Lightly moisten fiber before making cordage.

Reverse Twist Cordage

Choose a bundle of fibers holding one end in your non-dominant hand.

1. With your dominant hand, roll the fibers between your thumb and index finger, twisting the fibers until they start to kink in the middle.

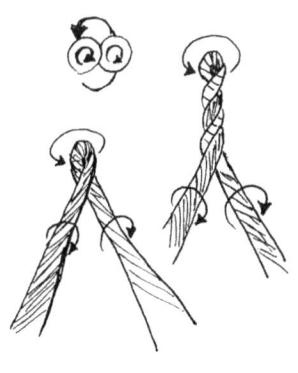

2. Pinch the kinked point with your non-dominant hand (index finger and thumb) so you have the two plies extending in a "Y" formation.

3. Pinch the top strand with your dominant hand and roll it away from your body then twist it down over

the other strand. Repeat: roll up, twist down. Repeat this step over and over until you have a length of string.

Good job! You can tie a knot in the end to keep it from unraveling. You can contine to make any length of string, and even double up lengths to make stronger string.

Looped Bag from Foraged Fiber

Once you have mastered the steady, repetitive nature of preparing and twining cordage, it's a deeply satisfying process. You may want to move on to using your cordage to make something. A looped bag (a knotless netted bag) is a great way to use cordage and bags are so versatile. Looped bags can be used for foraging, for purses, or to hold a water bottle, a phone/camera, or a wallet.

Follow the previous section for making cordage. Once you have an four or five feet made, you can tie a loose knot at the end and begin your bag as follows:

- Make about 4-6' of cordage. Leave the working ends loose, so you can easily splice in more length later.
- The looped weave begins with a foundation knot, a simple slip knot. Note the slip knot is 'reversed'; pulling on the tail end tightens the loop while the working end is stable. Put an extra overhand knot in the tail end to prevent it pulling out.
- Then thread a large needle, using the other, working end of your cordage.
- Make a simple half-hitch into your slip knot, using your thumb as a guide for the size of the loop. Make 6 half-

hitches into the slip knot, and then pull the slip knot closed and tie a knot in the loose end to keep it closed.

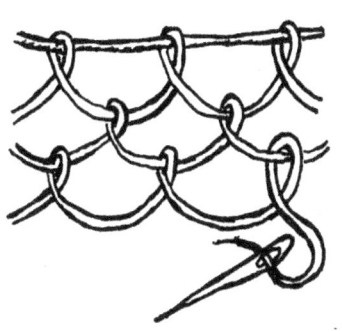

- Now step up and make two half-hitches in each loop of our first round. Once you complete the second round lay your work flat and see if you want to continue increasing the size of the circle. If so, each row, put two half-hitches in six loops of the previous row.
- As weaving continues and the cordage gets to short to work with, you can get some more fiber and make another 4-6 feet of cordage. Then you can return to weaving it in.
- Continue to build your circle base until it is as wide as you want your bag. At that point, you can make a single half-hitch in each loop to make the sides of the bag.
- When the bag is as tall as you want it, tie off the cordage. You can string a separate length of cordage through the top loops of the bag to make a drawstring.

Computers

Computers are time savers and time wasters. Computers can both save and use resources. So why is this a topic in our green book? Is there a way to "green" our use of technology? Some of us believe there is.

Open source software may help us green our computer use in several ways. Linux is the most popular open-source operating system for the general user (www.linux.org), one that I have been using for over a decade. This list will focus on the Linux project.

- Linux is free as in beer: you don't have to pay for it. It is also free as in speech: you can freely use, share and change it.

- Linux is less vulnerable to viruses and malware than other operating systems due to its software architecture and its smaller market share.

- Opensource is a development strategy that uses crowd-sourcing to gather and implement the best thinking on development and implementation. Many software developers depend on secrecy, keeping their ideas safe from theft so that they can make a profit. There is additional software added to programs and operating systems to protect them from you-the user. Opensource developers share their ideas and their source-code (the internal blueprint that shows how the software works). Thus opensource builds a community of open sharing.

- Linux is generally designed to run on older and smaller computers. This makes it run super-well on modern computers.

Another aspect of green computing is how we dispose of old equipment when we get something new. Some make a commitment to pass-on their old phone or laptop, when they get something new. They might give the device to a friend, relative or aquaintance who needs it, or give it to a nonprofit that restores and distributes them. Alternatively, they might sell them on craigslist or eBay. However it is done, there is a commitment that the device not end up in a landfill. And there is a further commitment to pass on the device in a timely manner, so that it is still something of value.

ENERGY

HOME ENERGY AUDIT

One of the missions of the Dogtown Ecovillage is to create a more ecologically sustainable neighborhood and a key component of this is decreasing our carbon footprint. Approximately 40% of an individual's carbon footprint is related to home electricity and gas consumption.

The lowest hanging fruit for most homes is to address areas where the house is leaking air and is under-insulated. The older brick homes in our neighborhood tend to leak around the plumbing stack, where the frame meets the roof and where the foundation meets the frame. Some houses also have very outdated heating, ventilation and air-conditioning (hvac) equipment. Professional home energy audits can help by pinpointing problems and addressing them intelligently. Well planned upgrades usually pay for themselves through lower bills within 3 years.

A home energy audit consists of a visual inspection of insulation and hvac equipment followed by a blower door test and thermal gun testing. A blower door test involves placing a piece of plastic and a fan across the opening made by the front door.

The fan then blows air out of the house and can measure how leaky the house is overall. By using an infrared heat gun (or by holding up one's hand to feel where the wind is coming from) the inspector can see where the leaks are coming from and where it would be best to seal up and place insulation.

COMMUNITY

Community is related to sustainability and to green living. Generally we can live more sustainably by working together, by sharing resources and by supporting a local economy and local food security. Ecovillages deliberately combine the notion of community and ecological sustainability. Often other forms of community living will have an interest in sustainability as more or less central to their way of being together.

Community and sustainability often walk hand in hand, and along the road we often find the value of justice as well. Many communities find their members joining in certain advocacy, for sustainability of course and also for economic justice, and for the rights of the poor and marginalized. One movement saw this as working for an "economically just, ecologically sustainable and spiritually meaningful world." These values and orientations are often found in abundance in community movements.

Ecovillage

Ecovillage is a term that covers a wide variety of sustainable communities that may be urban or rural, and that are generally committed to living in a more ecologically sustainable manner. They may be small or large, loosely or more tightly bound together, and may have more or less formal membership, governance and financing. On the continuum of each of these issues, the Dogtown Ecovillage is urban, it is loosely organized, geographically we are spread out in our larger neighborhood, but many of us can walk to each others' houses for potlucks, events, etc.

There are other similar types of arrangements. For example co-housing has gained some popularity. Co-housing is an arrangement where each family unit owns it's own space, and there are also shared spaces, like a kitchen, gathering space, gardens, offices, etc. This requires some initial investment in infrastructure and some criteria for financial commitment, for shared responsibilities in the community and often for the level of shared values in the community.

Co-ops may be cooperative housing that is jointly owned and occupied, a cooperative business venture, or a buying cooperative. These may be found as stand-alone entities, and are commonly found in conjunction with other forms of community described above. For example, our ecovillage could support a car co-op and we have done some planning for a food co-op. While the ecovillage would plan and establish these co-ops, we might invite participants from outside the ecovillage.

Potluck

Potlucks are a staple of intentional communities. Bring what you have to share, what you grew, what you cooked. Share it with one another. It's easy to plan, easy to prepare and you get to see what everyone else is growing. Most of our potluck dishes are vegetarian, but there's no hard fast rule on that for our group. And sometimes we have a good old fashion barbeque – and folks decide to participate or not.

Potluck shares the cost and the preparation time, and I am always amazed at the great food that we get. In our neighborhood where we can walk to each others' houses, you'll see folks converging with their bags or pots. I think it must be like an old fashion neighborhood block party. Only we make a regular thing of it. A few tips:

- One dish works well, a salad or cassarole.
- Make sure to bring your own serving bowl, serving spoon, etc. And make sure you bring them home. Sometimes I assemble a salad on site. I'll bring the greens, seasoning, quinoa and rice or pasta. Then I'll assemble it fresh.
- In a community group – it's helpful if someone has a set of dishes for everyone, and cloth napkins as well. For some groups, everyone brings their own dish, utensils and drinking cup. It makes it easier for the host.
- Try to avoid disposable dishes, etc. Let's go eco-friendly!
- I do regular potlucks with different groups, so I have a basket that fits whatever food I'm bringing and any extras I may have for the group.

Outreach

One aspect of community is the movement of members into and out of the community. Dogtown Ecovillage is a very loosely structured community, so we are open to new people joining us, and we have had people leave our group as their lives moved in different directions.

One aspect of our community and our planning is outreach to those who may be interested in joining us. Early on, we defined our goals, this helps to let people know who we are and to let people assess whether they would like to join us for occasionally, or more regularly.

Our primary outreach flyer has been a map of our houses, and a list of our goals. We have a facebook page and a email group that help us with internal communication. When someone is interested in joining the group, we usually direct them to those two places for more information about gatherings.

We have been listed on www.ic.org – the Fellowship of Intentional Communities website and we get occasional inquiries from that.

We are generally looking for like-minded people who already live in the neighborhood and would like to collaborate for sustainability and community-building. Some people learn about us, and then when they are moving, they look in our area, along with other areas. We weigh many factors in choosing where to live – those who may want to be part of the ecovillage would probably consider living in our Dogtown center as one of those factors.

We take part in various community and sustainability events across the city, both individually and as a group. We have taken part in the St Louis Earth Day festival for a few years, and that has been great for us as a community and has helped to get the word out about the ecovillage.

Since 2012, each year we have had a few families join, either those living here already, or those who move into the neighborhood. Some are more active than others. Some are more active in projects, others more in community events. We have a few hundred people in our facebook group and about one hundred in our email group. When it comes to potlucks or projects, we usually have anywhere from a handful to twenty people. I consider this a comfortable group. I know the people who come regularly, and we often have new people joining. They seem to fit right in.

Empowering

Our ecovillage works a lot on individual empowerment. We are a loosly knit community group that shares values of sustainability and community building. When someone in the group comes up with a plan, we generally talk about it and then, if there are enough people interested in making it happen – then they just go for it. The whole group doesn't have to sign off on every project.

This empowers people to take initiative and invite others to collaborate with their project. Some have tools, supplies, plants, or other things to contribute to the project. Others can contribute services.

Perhaps the person proposing a project initially thought the whole group would be involved. Because of who shows up though, they scale their goals to the response they receive. They may also adjust their project to meet additional goals and concerns of their newly recruited collaborators. In the end they have been empowered by the group, supported by the group and they do what is possible.

Consensus

Many communities and ecovillages try to work by consensus. This means many different things to many different people. Generally though, it means we don't just vote on somebody's idea. We listen to the various ideas that are proposed. Then we work to incorporate all the ideas and come up with a project or proposal that works for everyone.

Because we are a loosely structured community, it is somewhat easier to reach consensus. No one will be loosing their house or their livelihood due to a decision of the group. People who disagree can simply step aside from the project.

Our style of consensus requires us to be attentive to who is showing up. If we decide this is an awesome idea, but only three show up – then we may need to re-think what consensus means. Are there projects that would have greater involvement? Or is our three-some project just what we want to be doing?

The combination of empowering sub-groups to take on projects, and consensus about bigger projects seems to be working for us, with our style of ecovillage.

Economics

There are many forms of intentional community building. You can find a lot of information about the topic, as well as information about individual communities at www.ic.org. Some of the types of communities are as follows:

- Ecovillages – Communities with a strong ecological and sustainability focus, with varying living arrangements.
- Cohousing – Living arrangements with both private space and shared common spaces.
- Communes – Full sharing of living space and income.
- Co-ops – Cooperative arrangements that focus on one or more aspects of living, e.g. food, transportation.
- Faith-based Communities – One of the above types of communities that has a significant commitment to religion or spirituality

Economics is a major part of any discussion of community-building. How much of the economy is shared. Shared economy creates an interdependence within the group which has several effects, both positive and negative.

- It can help to build strong bonds in the group and a sense that all are in it together.
- It can be a barrier to entry, for those without means.
- It can be a sticking point in forming community because of the time and planning required to clarify needs and expectations of group members and find an economic model that works for the group.
- It can help keep people in the group, once they have done the hard work of creating and transitioning to the shared economy.

Most communities opt for some sort of partially shared economy. Some is formal some is informal.

- Potlucks are probably the simplest form of shared economy. Everyone brings something and they share an amazing meal.
- Tool sharing co-ops may be as simple as a Facebook page or an email list where people can
- Community gardens are another form of shared economy. In some, gardeners are restricted to their individual bed, though there is often sharing of tools, seeds and harvest. In other community gardens, all work together on the garden plan, on the planting and cultivation and all share in the produce.
- Housemates will require some shared economy. At a very least the group will have to budget expenses and collect regularly from each housemate.

One thing is clear to those who spend much time in and among communities. There is an inverse relationship between inclusivity and sustainability. A group may value inclusivity to the point that everyone who cares to show up is included in the group. But then group identity can suffer. More critically, if no commitment is required, the group can swell its numbers, but the phenomenon of free-loading becomes an issue. People come only for what they can receive, but are unwilling to contribute to the community. If a group wants to be sustainable, it needs some boundaries, and some 'barriers to entry.'

Our ecovillage started with no economic sharing, other than our potlucks. Over time, we have experimented with various common projects, co-ops and cottage industries.

Co-ops

We can build co-ops around anything: food, transport, tools, garden, cooking. There is the story of an old farmer, let's call him Al. He lent his axe to his neighbor Bill. Bill saw that the axe-handle was worn and replaced it. Later, Chuck asked to borrow Al's axe, Al sent him to Bill to get it. Chuck noticed the axe handle was new, but the axe head was old and rusty, so he replaced it. He brought the axe back to Al – now a completely new axe that Al didn't recognize. So it goes in an old farming community – everyone works together because everyone knows they all do better that way. Cooperatives enhance sustainability. Today, with a more transient society, we have to work harder to establish cooperatives and to make them work. All the co-ops have some basic elements:

- **Idea** – we would like to work together and share something for the benefit of all. The something we share is the basis of the co-op. For example, a cooking co-op may consist of five households that each agree to cook five meals once a week, and deliver them to each of the other houses. Voila! An idea.

- **Members** – who is in and who is out? In the cooking co-op, there are five members and that works out. What if there were more people? Would they make a second co-op? Or would we cook more food, but less often. It's important to clarify membership and expectations of members.

- **Rules** – Who makes decisions, how do we communicate, how to we accommodate changes in plans. Again with the cooking co-op – what happens when someone is out of town, or is sick? What if someone wants to join or to leave the group?

- **Money** – If there is money, or even time or equipment, then there should be some way of ensuring basic equity in the group. The larger the sums and the larger the co-op, the more the need for clarity in this regard.

- **Duration** - You'll need some nitty gritty like how long are you continuing: one month, six months, indefinitely? How do you evaluate the co-op and end it if it is not working? How do you resolve disputes that may arise? How do people join and leave the co-op?

It is great to do informal co-ops, but in our urban settings, it is often better to plan a little and save a lot of heart-ache.

Tool Sharing

Did you ever begin planning a project around the house or garden, only to realize that you're missing key tools to make it happen? So what do you do? Head out to the store? Abandon the project?

If you're in our ecovillage, you put out a call for the missing tool. Often there is someone in the group that has what you need. We have talked about a more formal tool sharing co-op, but that has not happened yet.

Tool-sharing co-ops are often started with a list of tools owned by community members. Lawn-mowers, line-trimmers, leaf-mulchers, snow-blowers, ladders, tarps are garden items that rarely see full time use. There are also drills, saws, etc. that could be shared.

People keep their tools at home – and someone starts a list of who has tools they would be willing to lend. The contact information is there and the individuals just contact each other to borrow something.

Other tool co-ops are more formal. Everyone has to make a commitment to the co-op to get tit started. Some may contribute tools, others money, still others organizing skills others contribute tool maintenance or storage. Tools are checked out and returned, with parameters from tools that come back late or damaged. Some even purchase insurance.

For us though, we just put out a call for a tool and people decided on a case by case basis whether they are willing to lend. The more we receive help from others lending tools, the more willing we are to lend our tools out to someone else. And we get to know who will be careful with the tools and return them promptly.

DOGTOWN ECOVILLAGE

Conclusion

This book grew out of a community experience of sharing information and skills to enhance the life and sustainability of our ecovillage and of those within our circle.

AUTHORS

Daniel Berg is a lifelong environmentalist and helped start the Dogtown Ecovillage. He is married, has 2 children and works as a physician at a federally qualified health center in St. Louis.

Amy Hereford focuses on planting her way to sustainability. She grows native plants, perennials and edibles and generally works on living lightly on the earth, simplifying and learning wild-crafting.

Zuzana Kocsisova is a doctoral student in genetics at Washington University in St. Louis who hikes, gardens, and studies reproductive aging in *C. elegans*.

Top Ten

This chapter will explore various sustainability topics in the form of "top-ten" lists.

Top Ten Native Plants for Shade

- Groundsel
- Wild Ginger
- Celandine Poppy
- Solomon's Seal
- Native Ferns
- Virginia Bluebells
- Columbine
- Wild Sweet William
- Golden Alexander
- Jacob's Ladder

Top Ten Native Wildflowers for Sun

- Goldenrod
- Coneflower
- Coreopsis
- Blazing Star
- Asters
- Milkweed
- Sunflower
- Bee Balm

Blue False Indigo   Poppy Mallow

Top Ten (Mostly) Native Fruits
Elderberry	Chokeberry
Blackberry	Black Raspberry
Pawpaw	Currant
Gooseberry	Strawberry
Blueberry	Apple

Top Ten Vegetables for Home Garden
Tomato	Beans
Lettuce	Spinach
Peas	Sweet Potatos
Radish	Turnip
Onions	Melons

Top Ten Native Trees
Redbud	American Linden
White Oak	Prairie Willow
Black Locust	Black Gum
Red Maple	Flowering Dogwood
Fringe Tree	Pecan

Top Ten Culinary Herbs to Grow

Oregano	Thyme
Chives	Garlic
Basil	Cilantro
Mint	Tarragon
Lavender	Sage

Top Ten Medicinal Herbs to Grow

Lavender	Plantain
Thyme	Garlic
Chamomile	Passion Flower
Feverfew	Mint
Echinacea	Wild Bergamot

Top Eight Animals for the Urban Garden

Chickens	Quail
Bees	Compost Worms
Rabbits	Minnows for Garden Pond
Ducks	Dwarf goats

Top Ten Plus Sustainability Tips for Renters

Wash clothes in cold water.

Use your window shades to keep heat in or out, depending on season.

Turn off all lights, appliances and electronics not in use.

Switch to LED light bulbs.

Cancel all junk-mail.

Recycle.

Find a neighbor who will take your kitchen compost.

Eat a plant-based diet.

Eat local and organic foods.

Take the stairs.

Take shorter showers.

Share electronics, tools.

Grow a few herbs or plants in a container.

Top Ten Home Energy Saving Tips

Wash clothes in cold water.

Install a programmable thermostat.

Use your window shades to keep heat in or out, depending on season.

Turn off all lights, appliances and electronics not in use.

Switch to LED light bulbs.

Look for the Energy Star label on appliances.

Use low-flow faucets and shower heads.

Clean or change filters regularly.

Reduce water heater temperature to 130° F, and wrap the water storage tank.

Seal air leaks and properly insulate.

Top Ten Garden Tools

Hand trowel and/or Soil knife

Garden spade

Bucket

Garden gloves

Wire Basket for harvesting and cleaning veggies

Watering can

Over-pass pruners

Pruning saw

Hoses – (including drip line and watering timer)

Kitchen knife and spoon dedicated for garden use.

Top Ten Ways to Reduce Global Warming

Refrigerant Management	Wind Turbines (Onshore)
Reduced Food Waste	Plant-Rich Diet
Tropical Forests	Educating Girls
Family Planning	Solar Farms
Silvopasture	Rooftop Solar

Top Ten Backyard Birds in the Midwest

American Robin	Goldfinch

Cardinal
Hummingbird
Downy Woodpecker
Northern Mockingbird

Bluejay
Black-capped Chicadee
Common Grackle
White-brested Nuthatch

Top Ten Backyard Butterflies in the Midwest

Monarch
Common Buckeye
Cabbage White
Painted Lady
Silver-spotted Skipper

Black Swallowtail
Tiger Swallowtail
Great Spangled Fritillary
Grey Hairstreak
Cloudless Sulphur

Top Ten Edible Weeds for Salads

Purslane
Wild Violets
Wild Garlic
Chickweed
Wild Amaranth

Lamb's Quarters
Dandelion
Plantain
Mallow
Wood Sorrel

Top Ten Ways to Reduce

Cloth shopping-bags
Use a water bottle
Smaller home size

Use a travel mug
Buy used
Keep only clothes you wear

Cancel junk-mail
Share books, tools, etc.
Avoid single-use plastics
Buy only what you need

Top Ten Ways to Reuse

Buy at Garage Sales
Free-cycle, Craigslist, etc
Rechargeable Batteries
Glass jars
Use both sides of paper
Sell at Garage Sales
Use your Public Library
Community swap
Melt old candles into new ones
Water plants with grey-water

Top Ten Items to Recycle

Plastic bottles
Office Paper
Glass bottles & jars
Electronics
Paper products
Aluminum cans
Junk mail
Cardboard
Tires
Newspaper

*Place recycle bins in your home and workplace & empty regularly.

Top Twelve Plants for Baskets and Fiber

Dogbane
Milkweed
Amsonia
Pipevine
Honeysuckle
Grasses
Brambles
Willow

Yucca   Ash
Poplar   Oak

Top Ten Ways to Increase Garden Biodiversity

Plant native trees and shrubs for birds.

Plant native wildflowers for butterflies.

Plant native plants for pollinators.

Supply a source of water for birds and pollinators.

Add a pile of logs, or leave an old stump.

Add nesting sites for birds, bats or pollinators.

Remove invasive non-native plants.

Continue adding diverse native plants.

Avoid pesticides and chemical fertilizers.

Take time to observe and enjoy your garden.

www.ingramcontent.com/pod-product-compliance
Lightning Source LLC
Chambersburg PA
CBHW052326220526
45472CB00001B/295